METEORITES AND THEIR PARENT PLANETS

METEORITES
and their parent planets

HARRY Y. McSWEEN, JR.
The University of Tennessee

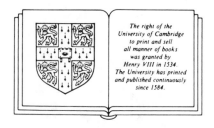

*The right of the
University of Cambridge
to print and sell
all manner of books
was granted by
Henry VIII in 1534.
The University has printed
and published continuously
since 1584.*

CAMBRIDGE UNIVERSITY PRESS

Cambridge

New York   New Rochelle   Melbourne   Sydney

Published by the Press Syndicate of the University of Cambridge
The Pitt Building, Trumpington Street, Cambridge CB2 1RP
32 East 57th Street, New York, NY 10022, USA
10 Stamford Road, Oakleigh, Melbourne 3166, Australia

First published 1987

Printed in the United States of America

Library of Congress Cataloging-in-Publication Data
McSween, Harry Y., Jr.
Meteorites and their parent planets.
Includes bibliographies.
1. Meteorites.   I. Title.
QB755.M465   1987   523.5'1   86–30996

British Library Cataloguing in Publication Data
McSween, Harry Y., Jr.
Meteorites and their parent planets.
1. Meteorites
I. Title.
523.5'1   QB755

ISBN 0 521 32431 9

TO MY MOTHER AND FATHER

Contents

Preface   *page* xi

1 Introduction to meteorites   1
From veneration to disbelief   1
The early history of meteoritics   3
Properties of meteorites   5
A fiery passage   13
Target earth   17
Frozen meteorites   23
Meteorite parent bodies   28
No one knows quite enough   31
Suggested readings   31

2 Chondrites   35
Once upon a time   35
Cosmic chemistry   41
The building blocks of planets   45
A recipe for chondrites   46
A fuzzy view of the early solar system   50
Reading the record   57
Raw materials for life   61
The miracle of creation   64
Suggested readings   64

3 Chondrite parent bodies   67
Meteorite orbits   67
Another way to look at asteroids   73
Structure of the asteroid belt   78
Sampling planetesimals   80
Asteroid heating   82
Chondritic dirt   88
Cosmic snowballs   92

Junk or treasure?	96
Suggested readings	98

4 Achondrites   101
  Origin and evolution of magma   102
  A geochemistry lesson   104
  The eucrite association   107
  The shergottite association   113
  Meteorites from the moon   120
  Ureilites   123
  Achondrite affiliations   125
  A personal touch   126
  Suggested readings   127

5 Achondrite parent bodies   129
  Our nearest neighbor   129
  A melted asteroid   137
  Looking for a needle in a haystack   140
  On a grander scale   143
  The red planet   147
  An inscrutable asteroid   152
  Melted clues   154
  Suggested readings   154

6 Iron and stony-iron meteorites   157
  The core of the problem   157
  Metal-loving elements   159
  Assembly directions for irons   161
  Order out of chaos   165
  Solidification of cores   171
  Cooling infernos   173
  Silicate inclusions   176
  Added complications   177
  Pallasites   179
  Mesosiderites   180
  Precious metals   182
  Suggested readings   183

7 Iron and stony-iron parent bodies   185
  Cores and raisins   185
  The core-mantle boundary   188
  A cornucopia of cores   190

Shiny beads 191
Asteroid families 195
Heavenly irons 198
Suggested readings 199

8 A space odyssey 201
Asteroidal traffic accidents 201
The properties of orbits 202
Geography of the asteroid belt 203
The planetary prison 206
Aten, Apollo, and Amor 208
Meteoroids exposed 211
At the finish line 214
An overview of meteorite history 216
On the importance of meteorites 218
Suggested readings 220

Appendix of minerals 222

Glossary 225

Index 235

Preface

Not so very long ago, the *earth sciences* were just that: the study of the earth. Not any more. Pick up any modern introductory geology textbook and you will find discussions of the moon and planets. Much of this new scientific turf is a legacy from astronomy, a result of the technological transformation of these bodies from dimly illuminated disks into crisply photographed worlds shaped by more or less familiar geologic processes.

The study of extraterrestrial materials, especially meteorites, does not appear to have been so readily assimilated into the earth sciences. I suggest that this is so because meteorites are commonly treated as if they were always small chunks of matter orbiting in space with no prior geologic histories. This book will employ a different tactic. My intent, as far as possible, is to trace meteorites back to their parent bodies, which are the sites of geologic processes. The bulk of this book is divided into a series of doublets – each consisting of a chapter describing certain related types of meteorites, followed by a chapter on their parent bodies. I hope this treatment will make these enigmatic objects more understandable.

Like any other science, the study of meteorites has its own vocabulary. I have attempted to minimize the introduction of new terms, but it is not possible to omit them entirely. Definitions of **bold** words in the text are given in a Glossary at the end of the book.

Any discussion of a subject this complicated may not always reach consensus; in such cases I have had to rely on my own prejudices. It is easy to read too much into meteorites, and in this book I may be more guilty of that than most. I also take responsibility for simplifying the subject by omission of some types of meteorites, when these do not provide insights into different concepts or processes.

Advances in meteorite research require the interplay of geology, chemistry, physics, and astronomy. This is a tall order, even for a subject that prides itself on its interdisciplinary (one might even say renaissance) character. I am indebted to researchers who are experts in these fields for their instructive discussions and review papers. In an introductory-level book such as this it is not possible to recognize individual researchers by name, although it would have been my preference to do so. I apologize in advance to any of my friends and colleagues in meteoritics who feel that their contributions have not been properly referenced. I acknowledge the following individuals and organizations for figures: Jack Berkeley, Doug Blanchard, Don Brownlee, Roy Clarke, Brian Mason, Ed Olsen, Robbie Score, and particularly the Smithsonian Institution and NASA. The cover and frontispiece art are the talented work of Mark Maxwell, who graciously allowed their reproduction. I am grateful to Peter-John Leone (Cambridge University Press) for being an enthusiastic advocate for this work, to Mike Lipschutz for a comprehensive and thoughtful review, and to Deborah Love, Melody Branch, and Denise Stansberry for their time and typing skills. Finally, I am indebted to my wife, Sue, and daughter, Lindsay, for patient understanding and for constant but welcome distractions, respectively.

Knoxville, Tennessee   H.Y.M.

1 Introduction to meteorites

In late 1492, three small caravels commanded by the Italian navigator Christopher Columbus boldly sailed into the New World. On November 16 of that year, the inhabitants of Cuba entertained these visiting sailors, who were no doubt luxuriating in the success of their epic voyage. Almost half a world away, the residents of the Alsatian (now French) village of Ensisheim also played host to a visitor, but this one had traveled much farther to make its appearance. On that day a stone weighing 127 kilograms streaked across the sky and impacted near the town. It must have caused quite a stir, as news of this unusual happening spread quickly. Upon hearing of the event, the emperor of the Holy Roman Empire, Maximilian I, under whose suzerainty the town fell, sought to capitalize on what he considered this omen of divine protection against the threat of Turkish invasion by ordering the stone to be placed in the local church. In accordance with medieval custom, the object was chained to the wall to prevent it from either wandering at night or departing in the violent manner by which it had arrived. This specimen, minus some fragments chipped off for museum collections in the ensuing centuries, still resides in Ensisheim today.

FROM VENERATION TO DISBELIEF

The Ensisheim incident was, of course, not the first account of the fall of a meteorite. Ancient chronicles from China and Greece document two independent meteoritic events occurring circa 650 BC, and earlier, though less definite, records of meteorite falls from Crete extend as far back as 1478 BC. The fall of a rock from the sky naturally was (and still is) a dramatic occurrence, and it is understandable that such events were described in detail and that the recovered objects were venerated. The Suga Jinja Shinto shrine in Nogata, Japan, has kept one fist-sized meteorite as a treasure of the religious center for more than a thousand years. Its date of fall

1

Fig. 1.1. Meteorite falls were commonly given religious significance by an-cient peoples. This woodcut carved in medieval times shows the fall of a meteo-rite near the town of Ensisheim (now in France) in 1492. The meteorite is illus-trated breaking through the clouds and also lying in a wheat field outside of the town. Upon recovery, the 127-kilogram stone was placed in the local church, where, except for some pieces removed for museum collections, it can still be seen today. Like many important events of this time, the fall and its effect on the local populace were immortalized in verse. Except for one other meteorite in a Shinto shrine in Japan, the Ensisheim stone is the oldest preserved meteorite that was actually observed to fall.

— May 19, 861 AD — is recorded in old literature as well as on the lid of the ancient wooden box in which it has been stored. A Rus-sian meteorite fall in 1584 was apparently memorialized by re-naming the town near which it fell (Tashatkan, literally "stony arrow"). The ancient Greeks and Romans enshrined and wor-shiped meteorites, and according to Titus Livius, one was even conveyed in a royal procession from its impact site to Rome, where it was revered for another 500 years. These objects were also val-ued by the Egyptians and have been discovered entombed with the pharaohs in pyramids. Prehistoric American Indians trans-ported meteorites long distances and sometimes buried them in crypts. One meteorite discovered in a burial ground of the Mon-tezuma Indians in Casas Grande, Mexico, was wrapped like a mummy. Even in more modern times, meteorites are sometimes accorded religious significance. The sacred black Kaaba Stone, to which Moslems in Mecca pay homage, is reported to have fallen from the sky, and some believe it to be meteoritic. Hindu religious literature states that meteorite falls herald important events, and in India it is reported that representatives of the Geological Survey must hurry to the site of any observed fall if they wish to collect the meteorite before it is enshrined by the local citizenry.

Most ancient philosophers viewed meteorites as heavenly bodies that had somehow been freed from their celestial moorings and had tumbled to earth. This explanation was not universally accepted, however; Aristotle considered them to be atmospheric phenomena. In this regard he was prescient in expressing the view of scholars in succeeding centuries, many of whom contrived to explain meteorites by terrestrial processes. Here is a typical example by W. Bingley, written in 1796:

It is but a trite observation to say, that the clouds make frequent visits to the waters of the earth, from which they usually carry away large quantities of that element, and with it, no doubt, the substances (even with some of the fish) which form the beds. . . . It is self-evident, that the streams which ascend with the clouds are sometimes clear as crystal, at other times thick and muddy. When the latter is the case then it is that these substances may be concreted; and, by some extraordinary concussion in the atmosphere, return to the earth.

Others argued that meteorites were terrestrial rocks that had been struck by lightning, an explanation that spawned the popular name "thunderstones" for these objects.

Many scientists, however, discounted altogether the idea that stones could fall from the sky. After the fall of a meteorite was witnessed and described in a document notarized by the mayor and 300 citizens of Barbotan, France, in 1791, the noted scientist P. Berthollet lamented:

How sad it is that the entire municipality enters folk tales upon an official record, presenting them as something actually seen, while they cannot be explained by physics nor by anything reasonable.

Meteorites could not begin to attract serious scientific scrutiny until such attitudes were dispelled.

THE EARLY HISTORY OF METEORITICS

Meteoritics is the name given to the scientific study of meteorites. The father of this discipline was undoubtedly E. F. F. Chladni, a German physicist and lawyer. His pioneering contribution was a small, 63-page book published in Riga in 1794. In it he argued that meteorites, at least those composed mostly of iron metal, with which he was most familiar, were extraterrestrial in origin. Based on evidence for their intense heating and their dissimilarity to terrestrial rocks, as well as the implausibility of other explanations, Chladni proposed a relationship between such meteorites and atmospheric **fireballs.** He correctly surmised that air friction heated objects traveling at high speed through the atmosphere, producing an incandescent glow.

Chladni's idea that meteorites were extraterrestrial amounted to scientific heresy, and his well-reasoned arguments were apparently not very persuasive to his contemporaries. Resistance to this hypothesis lingered in part because of scientific conservatism, but also because most meteorites were stones rather than chunks of metal, and thus were at least superficially similar to terrestrial rocks. However, Chladni's timing was perfect. Almost in immediate answer to his critics, in 1795 a large stony meteorite fell in the village of Wold Cottage, England. This event was important in refuting other currently popular mechanisms for the formation of meteorites (such as lightning or condensation from clouds), because the fall occurred out of a clear, blue sky. A specimen of the Wold Cottage meteorite eventually reached a young but highly respected British chemist, E. C. Howard, who decided to perform a detailed analysis. This study, done in collaboration with J. L. de Bournon, a French mineralogist exiled in England after the revolution, resulted in one of the first precise descriptions of a stony meteorite. In 1802, Howard reported concentrations of the element nickel in small grains of metal that de Bournon had separated from this stone. Nickel had earlier been analyzed in the metal of iron meteorites. Chladni's logical arguments had begun to persuade a number of scientists that iron meteorites were extraterrestrial, and this chemical link between irons and stones cleared the way for the interpretation that all meteorites had similar origins. Howard, however, was cautious in interpreting his data, although others were not.

After these two pivotal studies, changes in the attitude of the scientific community began to occur very rapidly. In 1803, a number of eminent French scientists, convinced by reputable eyewitness accounts of meteorite falls and by their own confirmations of Howard's chemical findings, threw their prestige behind the proposition that meteorites were extraterrestrial in origin. Remaining skepticism was silenced several months later when the town of L'Aigle, France, was peppered by a shower of no less than 3,000 stones. The French minister of the interior commissioned J. B. Biot, a physicist and one of the youngest members of the French Academy of Sciences, to investigate this incident. His 1803 report is commonly considered to be the turning point in the recognition of the authenticity of meteorites as extraterrestrial objects. In contrast to previous, rather dry scientific reports on meteorites, Biot's paper was dramatic and exciting. Nevertheless, its impact was made possible by the careful research of his predecessors, and the importance accorded to this work in recent times may have been overestimated.

Within a decade of the appearance of Chladni's book, his hypothesis that meteorites were extraterrestrial won general acceptance, and the science of meteoritics was launched. This is not to imply, however, that resistance to this new idea had totally vanished. America's scientifically literate president, Thomas Jefferson, apparently remained unconvinced. Upon hearing of the Weston, Connecticut, meteorite fall reported by Yale professors B. Silliman and J. L. Kingsley in 1809, he is reported to have said, "It is easier to believe that Yankee professors would lie, than that stones would fall from heaven." (This may be an embroidered version of Jefferson's comment, as its source is a secondhand report made by a person not actually present at the dinner party at which the statement was allegedly made.)

From the beginning of the nineteenth century forward, meteoritics progressed steadily to an exacting and highly interdisciplinary field. Developments during the nineteenth and twentieth centuries are much more interesting than those that led to the birth of this science, and many of these will be explored in subsequent chapters of this book.

PROPERTIES OF METEORITES

A **meteoroid** is a natural object of up to approximately 100 meters in diameter that is orbiting in space. A **meteor** is the visual phenomenon associated with the passage of a meteoroid through the earth's atmosphere. A **meteorite** is a recovered fragment of a meteoroid that has survived transit through the earth's atmosphere. Meteorites are named for the geographic localities in which they fall or are found. As a consequence, meteoritics is endowed with a heritage of exotic place names that add to the mystery of these objects.

We have already alluded to the existence of several types of meteorites – irons and stones. **Iron meteorites** consist almost entirely of nickel-iron metal alloys, whereas stony meteorites are composed mostly of silicate minerals, although many also contain small metal grains. In a third category are **stony-iron meteo-**

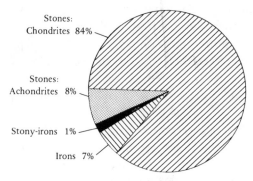

Stones: Chondrites 84%

Stones: Achondrites 8%

Stony-irons 1%

Irons 7%

Fig. 1.2. There are significant differences among fallen meteorites in the proportions of meteorite types. The shaded areas of this pie diagram illustrate the relative abundances of chondrites, achondrites, irons, and stony-irons. The proportions of the various classes of meteorites among finds are very different, being heavily weighted toward irons and stony-irons because their distinctive properties make them easily distinguishable from terrestrial rocks. The fall statistics more accurately reflect the proportions of meteoroids orbiting in the vicinity of the earth.

rites, which have nearly equal proportions of metals and silicates. Early and widely used classification schemes, such as those developed by the European petrologists G. Rose in 1863, G. Tschermak in 1883, and A. Brezina in 1904, referred to irons as "siderites," stony-irons as "siderolites," and stones as "aerolites." The stones can be further divided into two broad categories: chondrites and achondrites. A **chondrite** is a kind of cosmic sediment, an agglomeration of early solar system materials that has suffered little if any chemical change since its formation. In contrast, an **achondrite** is an igneous rock, the product of partial melting (accompanied by changes in chemical composition) and crystallization.

The proportions of iron and stony-iron meteorite **falls** (those seen to fall and then recovered) are very small, only about 7 percent and 1 percent, respectively, of the total number of meteorites collected. Of the remaining 92 percent stones, 84 percent are chondrites. These statistics may seem surprising to anyone who has looked at museum displays of meteorites, which are often dominated by irons and stony-irons. The ratio of irons to stones among meteorite **finds** (those recovered meteorites that were not observed to fall) is much larger than that among falls, because irons survive terrestrial weathering processes better and are more readily recognized as something unusual by non-geologists who stumble upon them. Irons also tend to be larger and more spectacular in appearance than stones, so museum exhibits are often biased

Fig. 1.3. Hoba, the world's largest meteorite, weighs approximately 55,000 kilograms. It still rests at its impact site in Namibia, where it is a national monument. This picture was taken in 1928 soon after the meteorite was discovered. As seen in the photograph, the flat area immediately surrounding the massive iron is a crust formed by terrestrial weathering, so the original meteorite was even larger. Photograph courtesy of Brian Mason (Smithsonian Institution).

toward them. The ratio of meteorite types in falls, as summarized in Figure 1.2, probably accurately reflects the proportions of objects reaching the earth from space, but any significance these statistics might have for objects in space is unclear. It would be more informative to have data on the relative masses of meteorite types, but even mass ratios can fluctuate over geologic time.

Meteorites come in all sizes, but there is a marked tendency for irons to be larger than stones. The biggest meteorite thus far discovered is Hoba (Namibia), a block of nickel-iron metal weighing 55,000 kilograms. Because it was proclaimed a national monument by the South African government in order to save it from the smelter, it still lies embedded in the limestone where it was first discovered. A photograph of this object, taken in 1928 soon after its discovery, is shown in Figure 1.3. Surrounding this meteorite is a layer of rusty weathered material that formed by terrestrial alteration of the outer part of the meteorite. If a correction is made for the amount of metal in this weathered halo, the original mass of this object becomes more than 73,000 kilograms.

In 1897 the American naval officer and Arctic explorer Robert E. Peary transported three massive pieces of another large iron meteorite from Cape York, Greenland, to New York City, where they can still be seen in the American Museum of Natural History. These, along with a fourth piece now in Denmark, have a collective weight of about 58,000 kilograms and together represent the second largest meteorite.

The largest meteorite found in the United States is the Willamette (Oregon) iron, weighing 12,700 kilograms. The meteorite was discovered in 1902 on property owned by the Oregon Iron and Steel Company, but the finder secretly moved the meteorite to his own property. This extraordinary feat, accomplished by mounting the meteorite on an ingenious trolley-like contraption drawn by horses, required three months of effort. The company successfully sued to repossess the meteorite after the finder began exhibiting it; this case established the legal precedent (at least in the United States) that a meteorite belongs to the owner of the land on which it falls.

There is a substantial list of iron meteorites weighing more than 4,000 kilograms, and two stony-irons are this large. In contrast, stony meteorites are rarely larger than 500 kilograms in weight. The largest stony meteorite known, a chondrite weighing 1,750 kilograms, fell as part of a shower of fragments in Jilin, China, in 1976. Some other stones have been found on the ground as groupings of related fragments that in a few cases may collectively weigh as much as a ton. Because they surely do not travel in space as closely grouped individual chunks, they must have been formed by disruption of larger masses on impact or during atmospheric transit.

The smallest meteorites are collectively called **micrometeorites** or, sometimes, **cosmic dust**. When the earth passes through an area where dust is concentrated, such microscopic particles produce meteor showers as they burn up during atmospheric entry. Some dust particles actually reach the earth's surface, but they are almost invariably melted. Miniature spherules collected in deep-sea sediments apparently formed in this way. An effort has been mounted to collect dust particles before they are destroyed by trapping them on sticky plastic plates located on the airfoils of high-altitude aircraft. Examples of a micrometeorite trapped in the upper atmosphere and one of its melted relatives recovered from the ocean floor are shown in Figure 1.4.

The earth sweeps up close to 10,000 tons of micrometeorite material each year as it orbits about the sun. The rate of fall for larger meteorites over the whole earth is more difficult to determine with any degree of certainty. One recent calculation, extrapolated from nine years of observations of meteoroids over Canada, suggests that 7,240 meteorites weighing more than 100 grams each fall every year (almost 20 per day) over the land area of the whole earth. This translates to one fall per square kilometer every hundred thousand years or so. Only a tiny fraction of this amount of material is actually recovered.

Meteorites that travel at high velocities through the atmosphere as single masses may develop distinctive shapes. Because a meteoroid has no shield such as those on spacecraft to dissipate the heat generated by atmospheric friction, its leading edge will melt. Ablation of this molten rock results in a smooth, featureless front surface. Some of the melt streams along the sides to the posterior surface, where it collects and solidifies into a roughly textured mass. The resulting conical shape is known as an **oriented meteorite.** If the shape of the projectile is not aerodynamically stable, it will rotate as it passes through the atmosphere. In this case its exterior will still melt, but no distinctive leading and trailing edges will be recognizable. A significant part (probably 30 to 60 percent of the mass) of most meteoroids is lost to melting and ablation in the atmosphere. Calculations of rates of ablation suggest that the anterior surfaces of typical meteoroids lose 1 to 4 millimeters of material during each second of flight time. As shown in Figure 1.6, the surfaces of some meteorites are marked with depressions resembling thumbprints (**regmaglypts**) that also form during atmospheric transit. These are probably caused by the violent motion of air or selective melting and ablation of certain parts of the meteorite. It is also common for incoming stones to break up into smaller pieces during atmospheric transit. Each of these will normally develop its own **fusion crust**, a layer of solidified melt **glass** coating the exterior. Such glassy surfaces are very thin, commonly less than a millimeter, except for solidified pools of melt on the trailing edges of oriented meteorites.

The external appearance of these objects produced by flight through the atmosphere often serves to identify them as meteorites. However, their internal compositions are also distinctive.

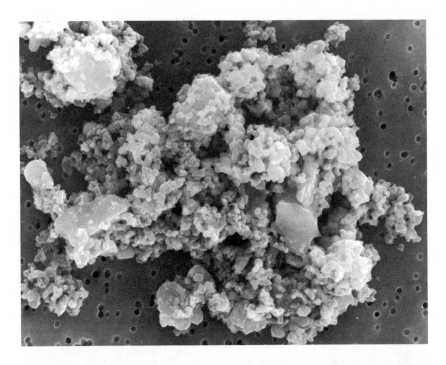

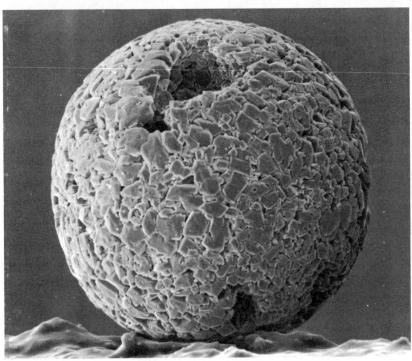

Fig. 1.5. *Partial melting of meteoroids due to friction as they pass through the atmosphere may produce objects with unusual shapes. The Bruno (Canada) iron meteorite shown here has six faces that were sculpted during atmospheric transit. The fine lines visible on the meteorite surface are frozen streams of melted fusion crust. Photograph courtesy of the Smithsonian Institution.*

Meteorites contain no new elements that are not already present in terrestrial rocks, but these are combined in some cases to form unusual compounds. The minerals schreibersite [(Fe,Ni)$_3$P], old-hamite (CaS), osbornite (TiN), and sinoite (Si$_2$N$_2$O), among others, have been recognized only in meteorites. (An appendix of minerals is presented at the end of this book.) However, these are relatively uncommon, and the major mineralogy of most meteorites is comfortingly familiar to geologists. Most of these minerals

Fig. 1.4. *The tiny micrometeorite above, viewed through an electron microscope, is only a few hundredths of a millimeter across. The fluffy aggregate of small crystals was collected in the upper atmosphere by U-2 aircraft. The spherule below, recovered from deep-sea sediments, is also only a fraction of a millimeter in diameter. It is a tiny droplet, presumably formed by melting of a micrometeorite like the one above during atmospheric transit. Photomicrographs courtesy of D. E. Brownlee (University of Washington).*

are silicates, such as olivine [$(Mg,Fe)_2SiO_4$] and pyroxene, a mixture of $MgSiO_3$, $FeSiO_3$, and $CaSiO_3$. Another common silicate that occurs both in meteorites and in terrestrial rocks is plagioclase, a feldspar whose composition varies between $NaAlSi_3O_8$ and $CaAl_2Si_2O_8$. The oxide minerals chromite ($FeCr_2O_4$) and magnetite (Fe_3O_4) are also common to both occurrences. The sulfide mineral troilite (FeS), cohenite (Fe_3C), and several forms of nickel-iron metal (kamacite and taenite) are common in meteorites but are extremely rare in terrestrial rocks and ores. The number of minerals that have been identified in meteorites by now probably approaches 100, but the great bulk of most meteorites is composed of the minerals already mentioned.

A FIERY PASSAGE

On the evening of November 8, 1982, a Wethersfield, Connecticut, family was watching television when they were jolted by an apparent explosion, later described as "like a truck coming through the front door." In actuality, a stony meteorite had punctured the roof of their home. The family quickly discovered a large hole extending up through their living room ceiling into the roof. Within a few minutes, firemen dispatched to the scene had discovered the intruder. The fire marshal later reported that the object was cool to the touch. The meteorite had plunged into the house at a 65-degree angle from the horizontal, finally coming to rest on the dining room floor.

One account of the Wethersfield fall was provided by an observer who was five miles away at the time:

It was close to 9:20 p.m., and I was jogging. I was headed straight west and saw a flash like lightning, and the entire sky was lit. I looked up, and about 5 degrees northwest of the zenith a very white and large object about the size of a basketball appeared. I thought it was space junk and saw fragments come off the eastern rim of it. The object stayed in one position and did not move — it was a head-on look. It could easily cast shadows. Six or more pieces fell from it in varying sizes. It then disappeared, and it never moved. I kept on jogging and began counting,

Fig. 1.6. The heat built up during rapid deceleration and the violent air movements around a falling meteorite may also produce thumbprint-like depressions called "regmaglypts." These two different views of the Allan Hills 81013 (Antarctica) meteorite illustrate the smooth appearance of the leading edge during flight (above) and regmaglypts on the trailing edge (below). Photographs courtesy of the Smithsonian Institution.

anticipating a sound so I could judge the distance. I counted between 30 and 50 seconds and heard a series of six or more "rifle shots" going off in the direction of Wethersfield.

Although meteorites do not usually crash into houses, what made this incident unusual, at least to odds-makers and the news media, was that another meteorite had fallen in the same small town just a few years earlier (also necessitating a roof repair). The second meteorite fell in the early evening, about 9:20 p.m. local time. In this regard it was also somewhat unusual, because twice as many fireballs are reported in the hours after midnight. This occurs because more are encountered in the direction in which the earth moves in its orbit, and the morning side faces the direction of the earth's motion. Other than these unusual circumstances, the 1982 Wethersfield fall appears to have been rather typical. Let us examine some of the phenomena associated with this fall more critically.

The maximum velocity of any meteoroid orbiting in the solar system is 42 kilometers per second, because objects with higher velocities would not have elliptical (closed) orbits. The earth's orbital velocity is 30 kilometers per second, so a head-on collision could potentially occur at up to 72 kilometers per second. This extreme velocity is very unlikely, however, because the meteoroid would have to travel in a retrograde orbit, moving in a direction contrary to practically all the bodies of the solar system. A meteoroid traveling in a prograde orbit should enter the atmosphere at substantially less than 42 kilometers per second, as the earth's velocity must be subtracted from that of a meteoroid overtaking the earth from behind. This is consistent with the velocities measured for most fireballs, which range from 10 to 30 kilometers per second. At altitudes below 100 kilometers or so the air density becomes great enough to create appreciable friction. This causes deceleration of meteoroids, as well as the melting and ablation we have already discussed. Air resistance eventually slows down most meteorites to a terminal velocity of a few hundred meters per second, at which point the body has lost all of its **cosmic velocity** (that which it had in space) and is free-falling under the influence of gravity. However, large meteoroids (those with weights exceeding about 10,000 kilograms) will not be affected as much by atmospheric resistance and can penetrate to the surface while still retaining at least part of their cosmic velocities. Besides its mass, a meteoroid's atmospheric deceleration will also be controlled by its entry angle and drag coefficient (controlled by size and shape).

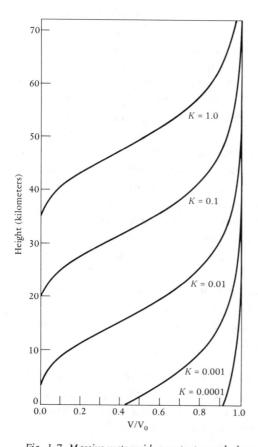

Fig. 1.7. *Massive meteoroids penetrate much deeper into the atmosphere than smaller ones before being slowed by air friction. The curves in this diagram illustrate the changes in velocity for meteoroids of various masses, shapes, and entry angles between the meteoroid's trajectory and the earth's surface. These variables are incorporated into the parameter K, which is a numerical value directly proportional to the drag coefficient (related to the shape of the meteoroid) and inversely proportional to the mass and sine of the angle of incidence. The horizontal scale gives the change in velocity (V) as a fraction of the initial velocity (V_0) at any height. Large meteoroids with low K values will penetrate the atmosphere without much loss of velocity, whereas small ones with K values of 0.01 or greater will lose all of their cosmic velocity at some point and finally free-fall to earth under gravity.*

Calculations of these effects on impact velocity are illustrated in Figure 1.7.

The jogger who witnessed the Wethersfield fireball reported that it was intensely white and that its glow was extinguished before it struck the ground. Some observed fireballs have been bright enough to illuminate 100,000 square kilometers of the earth's surface. The increase and subsequent decrease in emitted light from the Wethersfield meteoroid were consequences of its rapid deceleration and the subsequent reduction in air friction. At terminal velocity, meteoroids are no longer luminous. When a meteoroid penetrates the atmosphere at greater than supersonic speed, it creates a shock wave as the air in front of the object is compressed. This is an effect similar to that produced by a bullet or a jet aircraft. Such shock waves may generate sound phenomena resembling thunder or detonations, as noted at Wethersfield. Observers near the points of many meteorite impacts have reported consistent patterns of sounds: detonations analogous to those produced by supersonic aircraft, tearing or rumbling noises, and whistling or buzzing like that produced by falling bombs. These sounds are apparently produced by falling meteorites in the sequence cited, but they reach the observer in reverse order because of the time required for the sound to travel.

The 1982 Wethersfield meteoroid was seen to break up into six or more pieces during its flight. Nearly half of all fireballs are observed to fragment, most at altitudes between 12 and 30 kilometers. The fall of a disrupted meteoroid may result in a **strewn field** of many individual meteorites. In the normal case of an oblique approach, the pattern on the ground delimited by the meteorites will be elliptical. The largest masses tend to carry greater distances than their smaller relatives, because the former retain some of their cosmic velocity longer and thus travel farther. A sketch of one such strewn field is illustrated in Figure 1.8. This idealized shape can be modified, of course, by repeated fragmentations or other complicating factors.

The damage to the home caused by the impact of the grapefruit-sized Wethersfield meteorite was certainly not minor, but it was less than one might expect from a rock falling to the earth from space. Meteorites striking the ground may excavate small cavities, but typically they penetrate only to depths nearly equal to their diameters. In fact, many meteorites are found practically sitting on the surface of unconsolidated soil or snow. This is, of course, due to the appreciable deceleration produced by atmospheric friction. A falling meteorite is still dangerous, but there are no reported instances of anyone being killed in such an incident. There is a record of a young girl who was struck by a falling meteorite in Juashiki (Japan) in 1927, but she was not seriously hurt. In 1954, a woman was bruised when the falling Sylacauga (Alabama) meteorite crashed into her home, and the Nakhla (Egypt) achondrite that fell in 1911 reportedly hit and killed a dog.

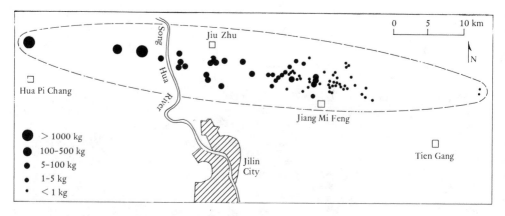

Fig. 1.8. Many meteoroids break up during atmospheric passage and fall to earth as showers of numerous fragments. In the case of an oblique approach, the impacting meteorites will form a strewn field of elliptical shape. This sketch map shows the strewn field for the Jilin chondritic meteorite shower that fell in Jilin Province (China) in 1976. Numerous specimens of various sizes were recovered; their relative masses are illustrated by location dots of different sizes. The shower apparently approached from the east, because the larger masses at the western end of the strewn field were more difficult to decelerate and tended to carry farther.

It may also seem surprising that the Wethersfield meteorite was cool to the touch, because a significant portion of it had to have been melted during its fall. Most of the molten material had already been lost to ablation, leaving only a very thin envelope of quenched fusion crust. Although in some cases fusion crusts may still be warm, the interiors of these objects certainly are not. Meteorites have been stored in the deep-freeze of space for eons, and atmospheric heating does not significantly affect their interiors, because heat conduction in stones or even irons takes much longer than the minute or so required for atmospheric transit. In one occurrence in Wisconsin, frost is reported to have condensed on the broken surface of a newly fallen meteorite, despite the hot July air temperature.

TARGET EARTH

We have previously noted that massive meteoroids entering the earth's atmosphere are not fully decelerated and retain some of their cosmic velocities all the way to the ground. Impacts of such large objects are fortunately rather rare events, because such bodies are uncommon and tend to fragment into smaller objects in the

Fig. 1.9. Although small meteoroids are decelerated because of atmospheric friction, they still may be traveling at considerable speed upon impact. The meteorite in the right hand of the lady in the polkadot dress penetrated the roof of her garage in Benld, Illinois, in 1938. It continued through the roof, back seat, and wooden floorboards of the automobile in the background, bounced off the muffler, and ultimately lodged in the seat cushion. Photograph courtesy of Edward Olsen (Field Museum of Natural History).

atmosphere. When such **hypervelocity impacts** do occur, they produce **explosion craters**. Massive meteoroids carry tremendous amounts of kinetic energy and have the potential to cause great devastation. For example, a stony meteoroid with a diameter of 2 kilometers and traveling at 50 kilometers per second would impact with an energy equivalent to 4 trillion tons of TNT. This nontrivial explosion would produce a crater 33 kilometers in diameter and 2.8 kilometers deep. This example is admittedly an extremely large meteoroid, but it is a reasonable size for some of the bodies that periodically are observed to cross the earth's orbital path. The last "close encounter" by one such object occurred in

1937, when it came within 750,000 kilometers (about twice the distance from the earth to the moon). Luckily, the chance that such a body might impact the earth any time soon is remote, but given the vast expanse of geologic time, it is not inconceivable that such an event has already happened. In fact, explosion craters have been found on the earth, but by all indications they were formed by impacts of significantly smaller meteoroids than that cited earlier. Meteoritic debris has been found at about a dozen such explosion craters. We have already seen that iron meteorites tend to be larger than stones, and it is probably no coincidence that all these craters are associated with irons and stony-irons.

There are about 70 craters on the earth that apparently were formed by hypervelocity impacts. Geologists have been slow to recognize explosion craters and, until the last few decades, somewhat reticent to accept hypervelocity impact as the mechanism by which they formed. However, explosion craters are not always as obvious as one might think. For example, many craters are filled with lakes and sediments. The New Quebec crater in Canada was not recognized until a prospector in 1950 noted the unusual circular shape of its contained lake on an aerial photograph. The 3.7-kilometer-diameter basin was not fully substantiated as a crater until scientists examined the area and the rocks carefully. The New Quebec crater is only 10,000 to 20,000 years old and is a relatively fresh feature. If it went unrecognized for so long, one can imagine that identification of older, more deeply eroded craters may be difficult indeed. The eroded remains of ancient craters are called **astroblemes**, literally "star wounds." Where weathering and erosion do not occur, for example on the moon, the surface is pockmarked with craters of all sizes accumulated over billions of years. On such bodies without atmospheres, even small incoming meteoroids can produce hypervelocity impacts.

There are numerous criteria other than associated meteorites that can be used to ascertain whether or not a circular depression on earth may be an explosion crater. The nature of the rocks in the walls and on the floor of the crater provide some clues. The cavity is commonly filled with shattered rock, and localized puddles of melted rock may be buried deep in the crater bottom. In a fresh crater, the surrounding rocks are bulged upward to form a raised rim, and in places original layers may even be folded back on themselves. The intense shock pressures in rocks outside the crater may form conical structures called shatter cones, the apexes of which point toward the center of the impact. Such high pressures also produce distinctive kinds of microscopic deformation in the target rocks, as well as transform certain minerals into more tightly packed crystal forms.

The cratering process can be visualized in three stages: compression, excavation, and modification. During compression, the kinetic energy of the meteorite is transferred to the target rocks in the form of compressional energy. The compressed rocks subsequently relax, causing crushed material to be excavated and propelled out of the crater. This material forms a blanket of deposited ejecta around the crater. Very large craters are subsequently modified in shape as rocks surrounding the crater fault or slump inward to produce central peaks or concentric rings. All of these stages may occur within seconds or minutes of the impact.

The best-studied example of an explosion crater is Meteor Crater, Arizona, pictured in Figure 1.10. A cross section of the crater is presented on the same page. This hole is about 1 kilometer in diameter and several hundred meters deep. However, it is now half filled with shattered and locally melted rock, and its original depth was greater. Rain has also washed other unconsolidated sediments into the basin. The rock **formations** into which the meteorite impacted have bulged upward to form the crater rim, and an overturned flap of sedimentary layers occurs on the eastern margin. Estimates of the age of the crater range from several thousand to 200,000 years, but a probable age of 50,000 years has been calculated from the extent of erosion of the rocks around the crater. Some of the sandstones in the target area contain coesite and stishovite, very dense minerals formed from original quartz grains subjected to high shock pressures. The surrounding desert is littered with fragments of the Canyon Diablo iron meteorite, which in aggregate weigh more than 18,000 kilograms. Meteorite specimens found on the crater rim contain small diamonds, whereas those collected farther from the crater do not. The diamonds, also very dense minerals, were produced by shock in fragments that may have been spalled off the rear of the impacting projectile. Specimens from the surrounding plains presumably were smaller pieces broken off the meteoroid in the atmosphere, and thus these may have been decelerated and probably did not experience the extreme conditions of hypervelocity impact.

The original impacting body that formed Meteor Crater has been estimated to have weighed more than 10 million kilograms. The property owners formerly presumed that the main mass was buried in the crater, and lured by the possibility of retrieving signifi-

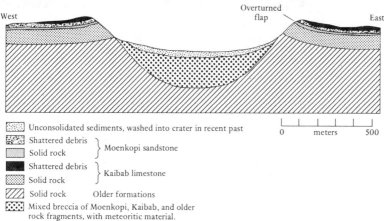

Fig. 1.10. Meteoroids that are so large that they are not slowed appreciably during passage through the atmosphere impart tremendous force when they strike the earth. These so-called hypervelocity impacts produce explosion craters, of which Meteor Crater, Arizona, is a classic example. The cross section (no vertical exaggeration) shows the stratigraphy of the crater. It is approximately 1.2 kilometers wide, and the depth to the top of the sediment fill inside is several hundred meters. The bottom of the crater is filled with shattered rock, and the layers of strata at the edge of the bowl have been overturned. Fresh explosion craters such as this are easy to identify, but older craters that have suffered erosion may go unrecognized. Holes were drilled by the mining company that owned this property in hopes of locating a large mass of buried iron meteorite, but they were unsuccessful. Photograph courtesy of the Smithsonian Institution.

cant quantities of nickel and platinum metals, they established a mining company. However, subsequent excavation and drilling, while helping to clarify the structure of the crater, did not strike meteorite. It now seems evident that in this and other large hypervelocity impacts, the projectiles are mostly demolished by fragmentation and even vaporization. Fine metallic globules that appear to be remnants of the original impacting mass have been recovered from the soil as far as 10 to 15 kilometers from Meteor Crater.

It has been proposed that large hypervelocity impacts may have been responsible for some of the great extinctions of living species that punctuate the geologic record. The best known of these extinctions marked the end of the Cretaceous Period some 65 million years ago and resulted in the demise of the dinosaurs as well as many other species. In all, about a fourth of the known families of animals disappeared over a short time interval. High concentrations of the elements iridium, gold, and osmium have been discovered in sediments deposited at the Cretaceous-Tertiary stratigraphic boundary in Italy, Denmark, Spain, and other locations around the world. These elements are greatly enriched in meteorites relative to terrestrial crustal rocks, and many scientists consider their high abundances to represent the geochemical signature of a massive meteorite. This chemical fingerprint apparently occurs worldwide in both continental and marine sediments of the same age. It seems possible that a large impact threw enough pulverized rock and dust into the atmosphere to block out sunlight, thereby lowering global temperatures and greatly restricting plant photosynthesis. Disruption of the food chain and/or alteration of climate could have snuffed out many land and sea creatures over a short period of time.

One of the problems with this idea is that none of the known explosion craters is of the right age or size. For this reason, some researchers favor an impact in the ocean. More than half of the ocean floor that existed 65 million years ago is now gone, having been pulled back into the earth's interior by inexorable plate tectonic movements and consequently obliterating the direct evidence of such an event. Although it seems possible that the Cretaceous-Tertiary extinction was caused by an impact, a number of other mass extinctions in the geologic record are not associated with iridium anomalies, and a few iridium anomalies (possibly representing other impacts) are not associated with extinctions.

FROZEN METEORITES

Antarctica is the coldest, windiest, highest, most arid, and most inaccessible continent in the world. It is at the same time a place of awesome natural beauty and glacial nastiness. On its rocky basement rests about 90 percent of the earth's ice, and within this is a treasure trove of meteorites. This unexpected source of extra-terrestrial material was uncovered by accident in 1969, when Japanese glacial geologists stumbled upon nine meteorites exposed on bare ice near the Yamato Mountains in Queen Maud Land. Thinking that the meteorites were pieces of the same fall, the scientists collected and sent them home for study. To everyone's amazement, the nine specimens included examples of four different classes of meteorites. Japanese geologists returned to the Yamato ice fields in 1973 specifically to search for meteorites, and in 1976 American field parties dedicated to this purpose began exploration of the interior flank of the Transantarctic Mountains. These groups have now collected thousands of meteorite specimens from a number of regions in the Antarctic, the locations of which are identified in Figure 1.11. The most productive areas have been the Yamato ice fields and the Allan Hills in Victoria Land.

In these locations vast swarms of meteorites are scattered over stretches of bare ice. The mystery of the manner by which meteorites are concentrated in these areas has now been at least partly unraveled. The polar ice sheet covering 12 million square kilometers provides an ideal catchment area for meteorites falling over long periods of time. These objects are frozen into the thickening ice and preserved in nearly pristine condition, protected from ordinary weathering by the frigid climate. This ice sheet reaches an altitude of about 4,000 meters, and in places its weight has depressed the underlying rocks to below sea level. The tremendous mass of the ice causes it to squeeze downward and flow outward toward the edges of the continent. The ice and enclosed meteorites creep along at rates of several meters per year, moving toward an ultimate fate of breaking off into the sea as icebergs. At some locations, however, the horizontal motion of the ice is arrested by obstructions. Effective barriers to impede the ice movement are provided by mountains (often not recognizable as such because they are nearly covered or even overrun by the icecap; the tips of mountains poking through the ice are called "nunataks"). Nearly stagnant ice is uplifted by forces pushing against the obstruction

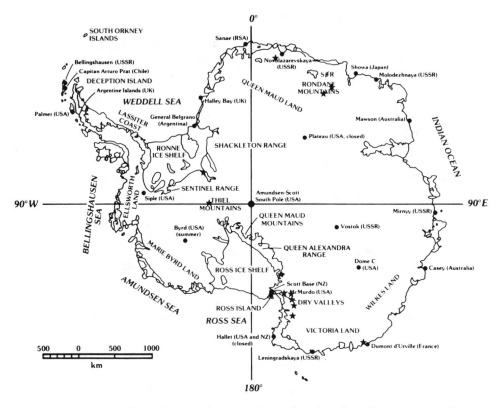

Fig. 1.11. Meteorite concentrations have been found in Antarctica in the Transantarctic Mountains near the Allan Hills (stars in Victoria Land) by USA scientific teams and in the Rondane Mountains near the Yamato Mountains (stars in Queen Maud Land) by Japanese expeditions. The individual meteorites recovered from these two areas in the last few years number in the thousands. Other areas such as the Thiel Mountains are expected to yield similar harvests of meteorites for future expeditions. Courtesy of NASA.

and eroded away by evaporation or ablation due to strong winds. These katabatic (descending) winds roar down the gently dipping ice sheet, clearing the stagnant areas of snow and producing a storm of dancing ice crystals that sandblast the ice surface. Measured ablation rates in such areas indicate that about 5 centimeters of ice are removed each year, continually exposing new ice at the surface. As each successive layer of ice, carrying an occasional meteorite, is uncovered and eroded away, the meteorite remains behind to join others and form a concentration. A diagram of the meteorite concentration process is illustrated in Figure 1.13.

Moving glacial ice is normally blanketed by a layer of compacted

Fig. 1.12. The dark color of meteorites provides a stark contrast to the Antarctic ice and snow and makes collection relatively easy. This chondritic meteorite was found by the author near Reckling Peak (Antarctica) in 1981. The stone is covered with dark fusion crust, but the lighter interior can be seen in the lower right corner where the meteorite has been chipped. The counter shows the number assigned to this sample in the field. Photograph courtesy of NASA.

snow, called "firn," that appears white. In contrast, stagnant ice that is being actively ablated is referred to as "blue ice" because of its distinctive coloration. Areas of exposed blue ice can be spotted in photographs taken by satellites orbiting in space, and this is where the search for Antarctic meteorites begins. Scientific teams are airlifted to blue-ice fields already identified in satellite photographs, and detailed searches are carried out on the surface by snowmobile. The collection efficiency at some localities has been remarkable, because the dark fusion crusts of even thumbnail-sized meteorites provide a stark contrast to the bright icy background.

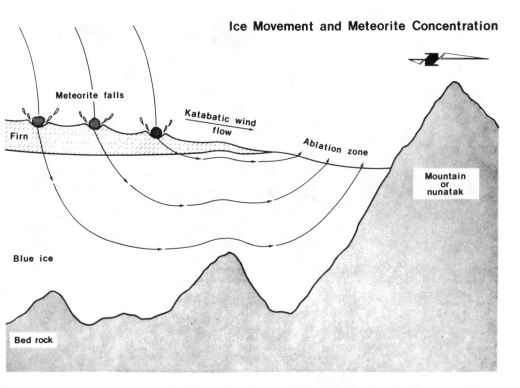

Ice Movement and Meteorite Concentration

Fig. 1.13. This diagram schematically illustrates the meteorite concentration mechanism in the Allan Hills and possibly other locations in Antarctica. All over the continent, meteorites fall into firn and are ultimately frozen into the accumulating blue ice. The ice sheet flows outward toward the edges of the continent unless it meets an obstruction such as a nunatak. Stagnant ice behind such a barrier will undergo ablation by katabatic winds, and meteorites will accumulate as successive layers of ice are exposed and removed in this way. Courtesy of NASA.

The meteorites are numbered, photographed, and collected in sterile plastic bags to prevent contamination of the samples. The Antarctic environment is clinically clean, and despite their residence in the enveloping ice, these specimens are less contaminated than most of those in museum collections. The still frozen meteorites are shipped back to the Johnson Space Center in Houston, Texas, where they are processed in a NASA laboratory used previously for the Apollo lunar samples. From this facility, processed samples are sent to research laboratories all over the world for scientific study. Eventually these meteorites will become part of the collection of the U.S. National Museum (Smithsonian Institution). Meteorites recovered by Japanese field teams are curated by the Jap-

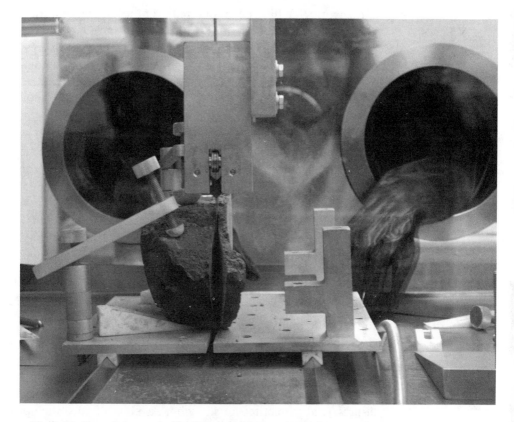

Fig. 1.14. Meteorites recovered by USA expeditions to Antarctica are trans-ported in their frozen state to the NASA Johnson Space Center in Houston, Texas. The meteorite processing laboratory provides curation and processing fa-cilities similar to those accorded lunar samples returned by the Apollo missions. Here a technician cuts a slab through a large achondrite. The saw is enclosed within a glovebox that preserves a contamination-free environment. Photo-graph courtesy of NASA.

anese National Institute of Polar Research in Tokyo. It is difficult to know how many of the individual meteorite specimens collected so far in Antarctica represent individual falls. By 1985 the number of specimens had grown to more than 7,000. Clearly, many of these must be fragments of the same meteorites, but the concentration mechanism has made it difficult to use field relations to ascertain whether or not individual samples are related. If only 5 to 10 percent of the samples so far recovered represent individual falls, then at least 700 to 1,400 new meteorites have already been added to our collections in less than two decades. This compares with approximately 2,600 catalogued meteorites in museums all over the world that have been collected over the last two centuries.

These meteorites provide a difficult problem in nomenclature. Because of the high concentrations of meteorites within small areas and the scarcity of local geographic landmarks, the naming system used for other meteorites is not applicable, and a special system has been devised. An Antarctic meteorite is identified by some nearby landmark such as the Allan Hills (normally abbreviated in capital letters, as in ALH for this location), plus a number such as 79014 (the first two digits correspond to the year in which the expedition that found the sample arrived in Antarctica, and the last three digits identify the specific sample in that year's collection).

Antarctic meteorites are especially interesting because they may have had long residence times on earth. By using sophisticated radioactive clocks in these specimens (the mechanics of which will be discussed in a later chapter), it is possible to determine the elapsed times since they fell. Ages measured in this way for these meteorites range from a few thousand to about 700,000 years; compared with Antarctic meteorites, even the Ensisheim fall was only yesterday. These recovered objects thus provide some perspective on the kinds and relative abundances of meteoroids that were orbiting in space and fell to earth before human history.

The Antarctic source not only has provided a large number of well-preserved meteorites for study but also has afforded the opportunity of discovering new types of meteorites. A number of rare or even unique kinds of meteorites have been recovered, further adding to the excitement of opening this meteoritic treasure chest.

METEORITE PARENT BODIES

The source of meteorites was (and is now) a riveting question for meteoriticists, and at least an interesting one to the general public. Early conjectures about meteorite **parent bodies** are as imaginative as those about the origins of meteorites themselves. In his ground- breaking contribution to meteoritics, E. F. F. Chladni recognized that meteorites did not form in space as the small chunks that periodically fall to earth, and he postulated that they must have been derived from the breakup of larger bodies "due to an external impact or an internal explosion." His ideas were already controversial enough that he wisely refrained from speculating on which heavenly bodies these were.

Well before Chladni's ideas were published, Titius von Wittenburg in 1766 had already noted a pattern in the spacings between planets. This sequence is a geometric progression, commonly expressed as

$$r = 0.4 + 0.3 \times 2^n,$$

where r is the radius of the orbit of any planet whose numerical order outward from the sun is n. Astronomers considered this regularity in planetary locations to be an important, primary feature of the solar system, especially after Sir William Herschel discovered the planet Uranus in 1781 at an orbital distance that closely fitted this equation. Using this rule, J. E. Bode forcefully argued that a planet was missing between Mars and Jupiter, and the relationship became known as the Titius-Bode law.

In 1801 the Sicilian astronomer G. Piazzi accidentally discovered a small planetary object located approximately at the Titius-Bode distance for the missing planet. Excitement coursed through the scientific establishment as Bode and others asserted that Piazzi's observation filled this planetary gap. In reality, Piazzi had discovered the first asteroid, 1 Ceres. However, the next year, H. W. Olbers discovered yet another body, the asteroid 2 Pallas, in the same orbital vicinity. Olbers was profoundly influenced by Chladni's hypothesis that meteorites are generated by the disruption of larger parent bodies, and he logically concluded that asteroids represent the fragmental remains of the missing planet in the Titius-Bode gap, as well as the source for meteorites. This idea became very popular and remained so for a long time. In the early twentieth century, there was no shortage of advocates for the position that all meteorites could have been derived from one large, stratified planet. In 1943, Harvard geologist R. A. Daly even provided a detailed reconstruction of the lost planet, which by that time had been christened "Phaeton," after the mythological character who rode a fiery chariot across the sky. Daly surmised that iron meteorites formed the core of this 6,000-kilometer-diameter planet, succeeded outward by a mantle of stony-irons and a crust of stony meteorite material.

However, a theory for the origin of the solar system formulated the following year by the respected Soviet academician O. J. Schmidt was not consistent with this exploded planet model. Instead, Schmidt suggested that the asteroids represented an arrested stage of planetary formation and had never been assembled into a large planet. This idea neatly circumvented the thorny problem of how to disintegrate a planet.

Several other variations of the small-parent-body theme also gained adherents at one time or another. In 1910 the astronomer G. V. Schiaparelli, most remembered as the discoverer of Martian "canals," determined a close orbital relationship between meteor showers and some comets, and he proposed that meteorites were cometary debris. Two Russian scientists, I. S. Astopovich and V. I. Vernadskii, even considered the possibility that meteorites came from outside the solar system, having wandered in from the surrounding galaxy. In a later paper, Vernadskii dismissed the asteroidal model as "an assumption based on seventeenth-century ideas, alien to celestial mechanics and universal contemporary views." If the history of meteoritics tells us anything, it is to be cautious in making definitive statements about meteorite origins.

The idea of small objects (whether asteroids or comets) as the sources for meteorites, although appealing to many, was not the only model for a meteorite parent body. A number of Chladni's contemporaries, including the physicist S. D. Poisson and the mathematician P. S. Laplace, both leading scientists of their time, advanced the hypothesis that meteorites came from the moon. They believed that meteorites were ejected from volcanoes, an idea possibly encouraged by alleged observations of active lunar volcanism by Sir William Herschel in 1787. This model remained popular until it later became clear that the moon had no active volcanoes. The lunar source was later resurrected by H. C. Urey, an American Nobel laureate in chemistry, who suggested that some meteorites were spalled off the moon by impacts. The idea was abandoned again when lunar samples returned during the Apollo program turned out to be distinct from meteorites known at that time. Urey has been vindicated, however, by some exciting new discoveries in Antarctica.

The hypothesis that meteorites might be derived from a large planet such as Mars appears to be a recent suggestion. The difficulties encountered in extracting a rock from a planetary gravitation field are formidable, but possibly not prohibitive.

Identification of the sources of meteorites remains today a contentious subject, but new observational techniques have been brought to bear on the problem. Moreover, the meteorites themselves carry information that is useful in reconstructing the histories of their parent bodies. To say too much at this point would give the story line away, but evidence bearing on the identities of meteorite parent bodies will be examined in later chapters.

NO ONE KNOWS QUITE ENOUGH

This chapter opened with a description of the fall of the Ensisheim meteorite in 1492. After the townspeople had secured this extra-terrestrial visitor in their church, they placed the following inscription near it: "Many know much about this stone, everyone knows something, but no one knows quite enough." Despite the fact that we have learned a great deal about meteorites in the ensuing five centuries, the veracity of this inscription remains unchanged. That, of course, is part of what makes this science perpetually interesting. The remainder of this book is devoted to explaining some of what we do know (or presently surmise) about meteorites. In the following chapters we shall consider each of the main classes of meteorites separately, focusing on what has been learned about their properties, origin, and subsequent evolutionary history. For each meteorite group we shall also attempt to reconstruct and/or identify its parent body or bodies.

SUGGESTED READINGS

The first three references are very readable without sacrificing quality. The other references are somewhat more technical, but still highly informative, and should be digestible by most readers.

GENERAL

Wasson J. T. (1974) *Meteorites*, Springer-Verlag, Berlin, 316 pp. (A somewhat technical but very informative book devoted to meteorite classification and properties. The book also contains an excellent reference list for papers on meteorites prior to 1974.)

Wood J. A. (1968) *Meteorites and the Origin of Planets*, McGraw-Hill, New York, 117 pp. (A somewhat dated but still excellent nontechnical overview of the importance of meteorites.)

Wood J. A. (1979) *The Solar System*, Prentice-Hall, Englewood Cliffs, NJ, 196 pp. (Chapter 5 gives a concise overview of the field of meteoritics.)

Hutchinson R. (1983) *The Search for Our Beginning: An Enquiry Based on Meteorite Research, Into the Origin of Our Planet and of Life*, Oxford University Press, 164 pp. (A lucidly written, nontechnical introduction to meteoritics.)

Dodd R. T. (1986) *Thunderstones and Shooting Stars. The Meaning of Meteorites,* Harvard University Press, 196 pp. (An engrossing, nontechnical account of meteoritics.)

CLASSIFICATION

Wasson J. T. (1985) *Meteorites: Their Record of Solar System History,* W. H. Freeman, New York, 288 pp. (An up-to-date technical monograph; Chapter 2 gives an excellent summary of meteorite classification schemes.)

HISTORY OF METEORITICS

Sears D. W. (1975) Sketches in the history of meteoritics. 1: The birth of the science. *Meteoritics* 10, 215–225. (Nontechnical paper describing a carefully researched account of the people and events that shaped meteoritics.)

Sears D. W. (1978) *The Nature and Origin of Meteorites,* Oxford University Press, 187 pp. (A concise technical book on meteoritics with an especially good section on its historical development.)

Marvin U. B. (1986) Meteorites, the moon and the history of geology. *Journal of Geological Education* 34, 140–165. (Nontechnical paper providing important details concerning the history of meteoritics.)

METEORITE RECOVERIES

Hughes D. W. (1981) Meteorite falls and finds: Some statistics. *Meteoritics* 16, 269–281. (Technical paper summarizing numerical data on meteorites.)

FALL PHENOMENA

McCrosky R. W. (1970) The Lost City meteorite fall. *Sky and Telescope* 39, 154–158. (Nontechnical account of the fall and recovery of the only meteorite photographed by the Smithsonian Prairie Network.)

IMPACT CRATERS

King E. A. (1976) *Space Geology,* Wiley, New York, 349 pp. (An introductory textbook with much useful information; Chapters 3 and 4 describe craters and impact processes.)

Silver L. T. and Schultz P. H. (1982) *Geological Implications of Impacts of Large Asteroids and Comets on the Earth,* Special Paper 180, Geological Society of America, 528 pp. (Technical proceedings of a conference devoted to evidence for a major impact causing geologic extinctions.)

ANTARCTIC METEORITES

Cassidy W. A. and Rancitell A. (1982) Antarctic meteorites. *American Scientist* 70, 156–164. (Nontechnical paper describing the mechanism by which Antarctic meteorites are concentrated.)

Fig. 2.1. Chondrites take their name from the small stony spherules, called chondrules, that they often contain in abundance. The origin of these small crystallized droplets remains one of the most perplexing aspects of chondritic meteorites. This photograph shows the sawed face of a chondrite found in Antarctica in 1977. This object contains an interesting assortment of rounded chondrules of various sizes. The small block is a scale measuring 1 centimeter on a side. Photograph courtesy of NASA.

2 Chondrites

Imagine a witness at the birth of the solar system, painstakingly observing and recording each event as it unfolds. What would such a recording be worth to science now? The origin and early evolution of the sun and planets are still, to a degree, shrouded in mystery, because there are no surviving witnesses or records – none, that is, but chondrites. The name of this important meteorite group derives from the ancient Greek word *chondros,* meaning "grain" or "seed," a reference to the appearance produced by numerous small, rounded inclusions called **chondrules**. Chondrules are visible in the photograph of a cut slab of an Antarctic chondrite in Figure 2.1. In this chapter we shall attempt to lift the shroud a bit and peek into the dark recesses of earliest solar system history. We shall do this by examining the record imprinted in chondrites. One might think that such chunks of rock would be mute witnesses, but nothing could be further from the truth.

ONCE UPON A TIME

If chondrites contain records of early solar system processes, they must be very old. But how old are they, and how does their age compare with that of the solar system?

The earth is our most accessible source of solar system material, and its formation age should be the same as that for the whole system. This, of course, presumes that there was no significant gap in time between the formation of the sun and the planets, and we have no theoretical reason or evidence to suggest such a hiatus. The only suitable clock for this determination employs naturally occurring isotopes. Atoms of a given element are distinguished from those of other elements on the basis of the numbers of protons in their nuclei, but different atoms of a given element may contain different numbers of neutrons. Atoms that differ in mass, that is, in the number of neutrons they contain, are called **isotopes**. Unstable (radioactive) isotopes transform, by loss of protons, neutrons, and electrons, over time into more stable isotopes of other elements. The parent isotopes decay at fixed (and measurable) rates, allowing the age of any sample to be determined from analysis of the amount of parent isotope that has decayed or daughter isotope that has formed. Unfortunately, these radioactive clocks are reset by geologic events that cause heating. This situation is analogous to a tape recorder that automatically erases the existing tape each time a new program is recorded. The radioactive clocks in terrestrial rocks are still suitable tools for geologic work, but they have been reset so many times that the formation age of the earth cannot be determined from them directly. In a few places, such as parts of Greenland and South Africa, are found the most ancient rocks thus far discovered on the earth, ranging in age up to about 3.8 billion years. These are very old, to be sure, but not old enough. The radioactive dates that these rocks record are metamorphic events, so the earth must be older still. Actually, determining the age of the earth is not as intractable a problem as it might seem, but in order to resolve this we must first determine the age of chondrites.

To illustrate the principle of a radioactive clock applicable to meteorites, we shall consider the decay of an unstable isotope (**radionuclide**) of the element rubidium, ^{87}Rb. The number 87 is the mass number of this isotope, equal to the sum of protons plus neutrons. Half of the unstable ^{87}Rb in any sample will decay to ^{87}Sr, an isotope of strontium, in about 49 billion years. This time interval is the **half-life** of ^{87}Rb. A knowledge of this rate of decay, along with measurement of the amount of ^{87}Rb that has been transformed into the new daughter isotope, will permit calculation of the age of this sample. The problem is a little more complicated, however, because not all of the ^{87}Sr now in the meteorite was produced by decay of ^{87}Rb — some was already there when the meteorite formed. The mathematical expression for this is

$$^{87}Sr_{now} = {}^{87}Sr_{original} + ({}^{87}Rb_{original} - {}^{87}Rb_{now}), \qquad (2.1)$$

where the terms in parentheses, corresponding to the amount of ^{87}Rb that has decayed, of course equal the amount of new ^{87}Sr produced.

The decay law states that the terms in parentheses, the original and final amounts of the unstable isotope, are related by $(e^{\lambda t})$. The term e is the number used as the base for natural logarithms (approximately 2.718), and in this expression it is raised to a power equal to the product of the rate of decay λ and the elapsed time t

(which is what we want to know). This decay law can be written as

$$^{87}Rb_{original} = {}^{87}Rb_{now}(e^{\lambda t}). \qquad (2.2)$$

Substituting this expression for $^{87}Rb_{original}$ into equation (2.1) gives

$$^{87}Sr_{now} = {}^{87}Sr_{original} + {}^{87}Rb_{now}\,(e^{\lambda t} - 1). \qquad (2.3)$$

^{87}Sr is not the only isotope of strontium present in the meteorite. ^{86}Sr also occurs, but it is not a decay product, and its amount does not change with time. However, because the amount of ^{87}Sr increases as ^{87}Rb decays, the ratio $^{87}Sr/^{86}Sr$ increases as time passes. Dividing both sides of equation (2.3) by a constant, ^{86}Sr, will not affect the equality:

$$\frac{{}^{87}Sr_{now}}{{}^{86}Sr} = \frac{{}^{87}Sr_{original}}{{}^{86}Sr} + \frac{{}^{87}Rb_{now}}{{}^{86}Sr}\,(e^{\lambda t} - 1). \qquad (2.4)$$

Notice that we can measure two of the foregoing ratios in the meteorite, $^{87}Sr_{now}/^{86}Sr$ and $^{87}Rb_{now}/^{86}Sr$. Typically, several separated fractions of the meteorite are analyzed, each fraction containing different minerals or at least different proportions of the same minerals. If we construct a graph plotting the two analyzed ratios for these mineral fractions versus each other, they will form a straight line like that in Figure 2.2. Because all samples on this straight line have the same age, the line is called an **isochron**. Equation (2.4) is the mathematical expression of this line, in the standard form $y = b + mx$, where m (the slope of the line) is $(e^{\lambda t} - 1)$, and b (the y intercept) is $^{87}Sr_{original}/^{86}Sr$, the initial strontium isotopic ratio.

Therefore, from the slope of the line in this diagram we can determine the age of the meteorite, and we have circumvented the problem of not being able to measure the amount of ^{87}Sr that was present in the original sample. As time passes, the slope of the line will become steeper. Isotopic data for mineral fractions of the Tieschitz (Czechoslovakia) chondrite analyzed by mass spectrometry[*]

---

[*] A *mass spectrometer* is used to measure isotopic abundances. A filament is carefully coated with the sample. When the filament is heated, the sample is vaporized, and individual isotopes are ionized. A large potential difference between the filament and a set of plates with narrow slits accelerates the charged ions through the slits, where they are collimated into a narrow beam. As this beam passes a powerful magnet, individual ions are deflected by different amounts, depending on their masses. The strength of the magnetic field is adjusted so that the path of the isotope of interest is focused onto an ion collector, which measures the beam intensity. The isotopic composition of the element is calculated by comparing the intensities for the different isotopes.

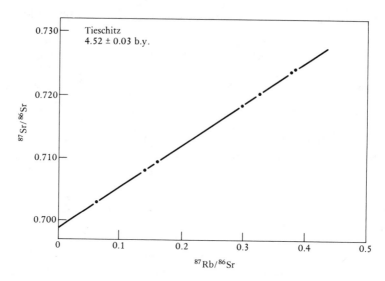

Fig. 2.2. *Precise measurement of radioactive isotopes provides a way to determine the age of a meteorite. In the method illustrated here, a radioactive isotope of rubidium (^{87}Rb) decays to a stable isotope of strontium (^{87}Sr), which mixes with already existing strontium isotopes (^{87}Sr and ^{86}Sr). The individual data points in this diagram represent minerals separated from the Tieschitz (Czechoslovakia) chondrite. The age of the meteorite (4.52 billion years) is calculated from the slope of the diagonal line, which steepens with increasing time.*

are shown in Figure 2.2. An age of 4.52 billion years was calculated from the slope of this isochron. Similar ages have been determined for other chondrites. Approximately 4.5 billion years ago rubidium and strontium were locked into minerals as they crystallized, and the radioactive clocks in chondrites started ticking.

But what is the relationship of this age to that of the solar system? What evidence is there to argue against the notion that chondrites may have formed later, say a billion years after the origin of the solar system? To answer this thorny question, we must turn to other isotopic systems. Unlike ^{87}Rb, some radioactive isotopes decay very rapidly. One example is an isotope of iodine, ^{129}I, which transforms to an isotope of xenon, ^{129}Xe, with a half-life of only about 16 million years. Any ^{129}I in the early solar system would persist at detectable levels for only 100 million years or so, equivalent to about six of its half-lives. Thus, if chondrites contain amounts of ^{129}Xe in excess of the normal solar system proportion of this isotope, they must have formed very soon after the now extinct parent isotope was created. Luckily, the atomic abundance of io-

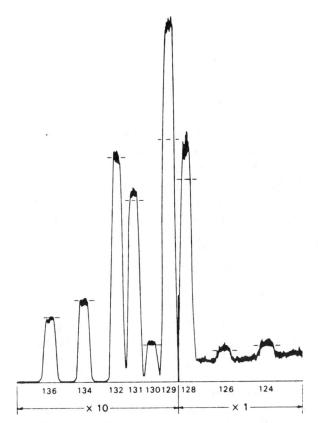

Fig. 2.3. Individual isotopes of a given element can be separated using a mass spectrometer. This diagram shows the proportions of xenon isotopes in the Richardton (North Dakota) chondrite. Peak heights correspond to the relative amounts of the various isotopes, and horizontal dashed lines are the proportions of different xenon isotopes in the terrestrial atmosphere, relative to ^{132}Xe. The large excess of ^{129}Xe in this meteorite formed from the rapid decay of an isotope of iodine (^{129}I). This evidence for now extinct radionuclides indicates that chondrites formed soon after the formation of the solar system.

dine is far greater than that of xenon, so enough excess ^{129}Xe might be formed by this process to alter the ratio of this isotope to other xenon isotopes perceptibly. Moreover, xenon is an element whose isotopic abundances can be measured with phenomenal precision. Figure 2.3 illustrates the mass spectrum of xenon isotopes in the Richardton (North Dakota) chondrite. For comparison, the dashed horizontal lines show the isotopic composition of xenon in the earth's atmosphere, probably a good approximation to the average ^{129}Xe content of the solar system. We know of no way in which

^{129}Xe in the earth's atmosphere could have been selectively removed relative to other xenon isotopes, so this comparison indicates that the ^{129}Xe excess in this chondrite is due to decay of ^{129}I.

Not all short-lived radionuclides produce daughter isotopes that can be directly measured. However, in some cases other kinds of fossil evidence of now extinct isotopes can be found in chondrites. For example, an isotope of plutonium, ^{244}Pu, with a half-life of only 82 million years, decays into an array of lighter isotopes. This process releases a great deal of energy, as is graphically illustrated by its unfortunate use in nuclear weapons. For plutonium atoms trapped in minerals, this energy is imparted to the newly created lighter isotopes, projecting them away through adjacent crystals at high velocities. These miniature projectiles leave trails of destruction, called **fission tracks**, that can be enlarged to microscopically visible sizes by etching the crystals with acid. Some other long-lived radionuclides, especially those of uranium, produce similar tracks, because they eject atoms with comparable energies. However, the densities of tracks in chondrites are too great to be accounted for by decay of uranium alone. The extra tracks point to the prior existence of ^{244}Pu when the meteorite crystals formed.

To summarize, chondrites formed about 4.5 billion years ago, as determined from slowly decaying isotopic systems. The former presence of short-lived and now extinct radionuclides indicates that these meteorites must have formed within at least the first 100 million years or so of solar system history, commonly called the **formation interval**. Therefore, we can accept an approximate age of 4.5 to 4.6 billion years for the birth of the solar system.

We began this discussion of time with a short discourse on why terrestrial rocks cannot be used to date the age of the earth, and by inference the solar system. Let us reconsider this question by examining uranium and lead isotopes. We have put this discussion off until last because the uranium-lead system is somewhat more complicated than the isotopic systems we have already considered, as there are several radionuclides of uranium (^{235}U and ^{238}U) that decay into different isotopes of lead (^{207}Pb and ^{206}Pb, respectively). By itself, neither of these parent-daughter pairs in terrestrial rocks can be used to date the earth's formation. However, this age can be determined by employing both systems. It is possible to model the evolution of both lead isotopes through time if we know the initial isotopic proportions of lead and uranium. The initial isotopic compositions can be approximated if we assume that the earth initially had the same isotopic ratios of uranium and lead that chondrites had. In 1956 a geochemist first made this assumption and calculated the age of the earth as 4.55 ± 0.07 billion years, essentially the same as that determined for chondrites. This may seem like circular reasoning, but remember that the time of chondrite formation has been determined from other isotopic systems.

COSMIC CHEMISTRY

The chemical composition of the solar system is often called the **cosmic abundance** of the elements. The solar system's composition is really equivalent to the chemistry of the sun, which contains most of the mass (more than 99 percent) of the whole system. The term cosmic abundance is therefore misleading, as it does not actually indicate the composition of the cosmos, but only a small part of it in which we are egocentrically interested. In fact, the sun has a different chemistry from other stars, and there is no way to estimate a truly representative cosmic composition.

The visible white-hot surface of the sun is called the photosphere. The chemistry of the photosphere has been measured by astronomers from the absorption of certain wavelengths of energy by elements in their excited states. The abundance of most elements in the sun has been determined to no better than ±40 percent of the amount present, but this level of precision is sufficient for our purpose.

Chemical studies of chondrites have a long and rich history. The famous chemist Antoine Lavoisier was a member of a commission of the French Academy of Sciences that performed the first crude chemical analysis of a chondrite, published in 1772. Modern research on chondrite chemistry has progressed to a level of high precision as ever more sophisticated techniques have been brought to bear on the problem. As a result, chondrites are probably the most thoroughly and accurately analyzed natural materials known, and the list of analyzed elements encompasses virtually the entire periodic table.

Figure 2.4 shows a plot of the abundance of each element in an average chondrite versus that element's abundance in the solar photosphere. These data are plotted on a logarithmic scale, with the numbers referring to exponents to the base 10. The abundances of different elements vary over many orders of magnitude, and a logarithmic scale allows all of these element concentrations to be plotted on the same diagram. The sun obviously contains much more of each element than a small meteorite, but ratios of

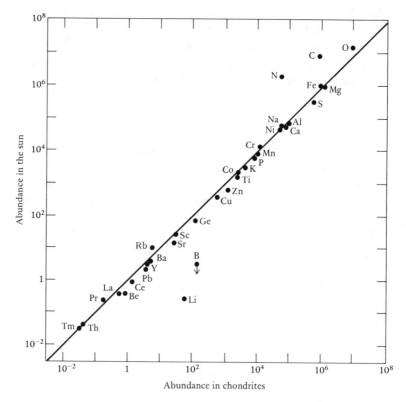

Fig. 2.4. *The chemical compositions of chondritic meteorites closely match that of the sun, suggesting that such meteorites represent primitive materials that have survived without significant change since the formation of the solar system. This diagram shows a comparison of the abundances of elements in carbonaceous chondrites relative to those in the sun. A perfect correspondence is indicated by the diagonal line. Both scales are logarithmic, and all elements are referred to 1 million atoms of silicon.*

different elements permit us to compare their compositions. In this figure, all measurements are referred to an arbitrary standard value, 1 million atoms of the element silicon. Exactly the same element ratios in the sun and chondrites would lie along the diagonal line. The correspondence between the two is very good, one might even say exceptional. No kind of terrestrial rock would show such correspondence, because the chemistry of rocks changes each time they undergo geologic processing. The major discrepancies involve the lightest elements (hydrogen, helium, carbon, nitrogen, oxygen), which are consistently more abundant in the sun. Under most conditions these so-called **volatile elements** exist primarily as gases. Thus, chondrites can be considered a sort of solar sludge, with compositions equivalent to the nonvolatile portion of the sun. Because all elements can be analyzed much more precisely in meteorites than in the sun, chondrite analyses are used to specify cosmic abundances of all but the most volatile elements.

Several other elements besides the volatile ones already mentioned also deviate from the diagonal line. Lithium (Li) and, to a lesser extent, boron (B) deviate the other way, that is, they are more abundant in chondrites than in the sun. These are real differences outside the limits of analytical error. The discrepancies can be explained by the fact that these elements are apparently utilized in fusion reactions that power the sun. Their solar abundances have been reduced during the past 4.5 billion years, so in this way chondrites actually record the chemistry of the ancient sun (hence primeval solar system) even better than does the present-day sun.

Several different **chemical classes** of chondrites with distinct compositions have now been recognized. This may seem like a contradiction. How can chondrites have solar compositions and yet be different from each other? Remember that Figure 2.4 is a logarithmic plot spanning many orders of magnitude, so concentration differences of a few percent are not really important. If meteorites were to be classified in the same manner as living organisms, we could say that different chondrites belong to different species, but all belong to the same genus and are distinct from all other taxonomic groups. The compositional classes of chondrites that we shall distinguish are the **ordinary chondrites** (so named because they are the most abundant type), the **carbonaceous chondrites** (actually misnamed when it was believed that they had higher carbon contents than other chondrites), and the **enstatite chondrites** (named for their high abundances of the mineral enstatite, $MgSiO_3$).

One of the most important differences among these chondrite classes is that they appear to have formed at different temperatures. Those that apparently formed at lower temperatures, the carbonaceous chondrites, contain a larger complement of volatile elements. Remember that no chondrite contains as high an abundance of volatile elements as does the sun. Figure 2.4 is actually a comparison between carbonaceous chondrites and the sun. Those that formed at higher temperatures, the ordinary and enstatite chondrites, are somewhat more depleted in volatile elements. In Figure 2.5, volatile elements are listed at the bottom of the figure

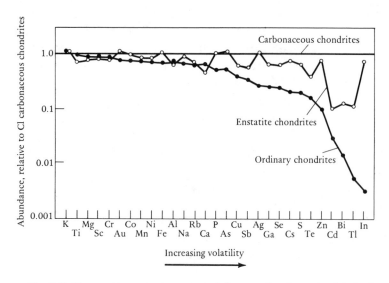

Fig. 2.5. The most important property of elements that determined their distribution in the early solar system was their volatility. Various chondrite classes appear to show different degrees of depletion of volatile elements. In this diagram, elements are plotted in terms of increasing volatility from left to right. Carbonaceous chondrites contain the highest proportion of volatile elements and presumably formed at the lowest temperatures. Relative to carbonaceous chondrites, enstatite and ordinary chondrites are depleted in volatile elements and must have formed at higher temperatures.

according to increasing volatility from left to right. Relative to carbonaceous chondrites, the other types of chondrites can be seen to have volatile element depletions that are dependent on volatility. Another important distinction between chondrite classes concerns their oxidation states. Oxidation causes iron atoms in the meteorite to become combined with higher proportions of oxygen. Almost all of the iron in carbonaceous chondrites is oxidized and is combined with oxygen into oxides and silicates. Some iron in ordinary chondrites is reduced and forms metal grains, whereas most of the iron in enstatite chondrites is reduced to metal.

Actually, these chondrite classes can be further divided into subclasses. The ordinary chondrites were first subdivided on the basis of distinctive iron contents into high-iron (H) and low-iron (L) groups in 1953. Subsequent analyses indicated the existence of a third group with even lower iron contents, called the LL group. Carbonaceous chondrites have been divided into similar subclasses, but these are not as well characterized because of limited numbers of specimens. The enstatite chondrites may also contain high-iron and low-iron groups.

What is the significance of all these classes and subclasses of chondrites with distinct chemical compositions and oxidation states? Some researchers think that there may have been a continuum of chondrite compositions in the early solar system, with different temperatures of formation and resulting depletion patterns of volatile elements controlled by distance from the sun. The density and temperature of the gas and dust from which chondrites formed presumably decreased with increasing solar distance, resulting in more oxidizing conditions and higher volatile contents. In this view, enstatite chondrites would have formed closest to the sun, and carbonaceous chondrites farthest away. Each distinct chondrite class or subclass that we observe today was presumably derived from one parent body. If this is correct, we have a very biased sample from only a small number of chondrite parent bodies.

THE BUILDING BLOCKS OF PLANETS

Thus far, we have learned that chondrites formed about 4.5 billion years ago and that this time represents approximately the beginning of the solar system. We have also established that chondrites have average solar system (cosmic) chemical compositions, at least for nonvolatile elements. These meteorites constitute the most ancient and chemically primitive kind of solar system matter. It does not require too great a leap of faith to infer from this that chondrites, or something very much like them, were the raw materials from which planets were assembled.

Current astrophysical models suggest that intermediate-sized bodies (**protoplanets**) grew by agglomeration of smaller bodies, probably of chondritic composition. Planets then formed by subsequent accretion of larger and larger protoplanets until most of the small debris in the solar system had been swept up or else ejected from the system. The gigantic impact basins on the surface of the moon and other planets are results of the heavy bombardment by protoplanets in the terminal stages of planet formation. Presumably the earth must also have looked like this at one time, but its impact scars have been erased by subsequent geologic activity.

If chondrites are leftover planetary building blocks, planets should have chondritic compositions. Unfortunately, the bulk composition of the earth, let alone any other planet, is difficult to estimate because initially homogeneous planets have now **differentiated** into cores, mantles, and crusts with different compositions. Except for a few chunks of altered mantle material that are brought up by ascending lavas, we obviously have samples only of crustal rocks for analysis. However, some elements are not easily separated from each other by geologic processes, and although the concentrations of these elements may change, their ratios will remain constant. The measured ratios of such elements are very similar in terrestrial rocks and in chondrites. As near as we can tell, the earth appears to have a nearly chondritic bulk composition.

A RECIPE FOR CHONDRITES

In his children's classic *One Fish Two Fish Red Fish Blue Fish*, Dr. Suess wrote:

> From there to here
> from here to there
> funny things
> are everywhere.

This little rhyme aptly sums up the situation facing the scientist who would catalog the components in chondrites. In addition to chondrules, chondrites consist of other funny (in the bizarre sense) things, including fine-grained matrix, metal grains, and irregular white inclusions, as illustrated in Figure 2.6. Different types of chondrites contain different proportions of these components, probably reflecting what was available in the neighborhood where each meteorite formed. The clumping together of these dissimilar materials is called **accretion**, and it was a very important process about which we know almost nothing. Each of these components probably contains a fossil record of some early solar system process or processes, but some are better (which is not to say correctly) understood than others.

There are several types of fine **matrix** materials that cement the chondrules and other components together. One type is mostly composed of tiny silicate grains of olivine and pyroxene, along with minor sulfides, oxides, feldspathoids, and clay minerals. The other type contains graphite, a form of carbon, mixed with an iron oxide, magnetite. To be truthful, there is no clear evidence of how either of these matrix materials formed.

Most chondrites are speckled with small pieces of metal. The metal consists of intergrown iron-nickel alloys with different com-

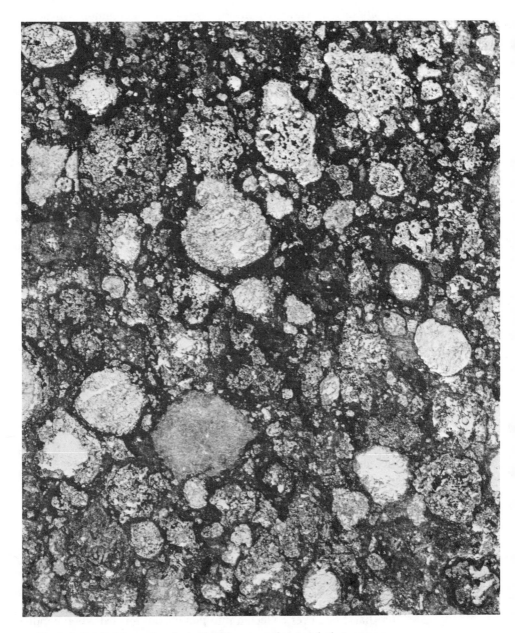

Fig. 2.6. Detailed inspection shows that the array of materials that compose chondritic meteorites is more complex than it might first appear. This microscopic view of a thin section of the Vigarano (Italy) carbonaceous chondrite shows numerous chondrules, white calcium-aluminum inclusions, and opaque metal grains, all held together by dark, fine-grained matrix material. All of this diversity is contained within several square centimeters of surface area in this meteorite.

positions and crystal structures. Two alloys are commonly present: taenite has a higher nickel content than kamacite, and the atoms in the former are more closely packed together. These alloys may also occur with iron-nickel sulfides and phosphides. Differences in the amounts of accreted metal partly explain the compositional distinction between H, L, and LL chondrites. Enstatite chondrites contain much more metal than other chondrite types, and carbonaceous chondrites contain the least.

Chondrules have always posed the most fascinating (and perplexing) aspect of chondrites. The German mineralogist G. Rose named chondrules in 1864, though references in the literature to "curious globules" in meteorites appeared as early as 1802. The first real breakthrough in understanding the origin of chondrules was made by H. C. Sorby, who began studying chondrites at about the time that Rose completed his work. Sorby was an English gentleman whose personal wealth enabled him to devote all of his energies to his consuming interests in science. His major contribution was the invention of the petrographic microscope, which today remains one of the basic tools of geologic research. By examining paper-thin slices of rock mounted on glass slides (called "thin sections"), Sorby was able to elucidate features otherwise impossible to observe. After years of studying thin sections of terrestrial rocks, he finally turned his attention to chondrites. From the microscopic textures he observed, Sorby concluded that these droplets had crystallized from molten material, and he drew comparisons with glassy blow-pipe beads and furnace slags. He also noted that these spheroids must have cooled prior to being accreted into chondrites.

... melted globules with well-defined outlines could not have formed in a mass of rock pressing against them on all sides, and I therefore argue that some at least of the constituent particles of meteorites were originally detached glassy globules, like drops of a fiery rain.

Recognizing that high temperatures would be required to melt chondrules, Sorby suggested that chondrites might be pieces of the sun ejected in solar prominences or might be residual cosmic matter than had never collected into planets but formed when conditions at the sun's surface extended farther out into the solar system. In the case of the first hypothesis, his views were undoubtedly colored by the concept prevailing at the time that the sun was a solid, rocky body wrapped in incandescent gases. His second hypothesis was surprisingly close to our current concepts.

In the intervening years, a number of alternative ideas to explain the origin of chondrules have emerged. They have been suggested to be droplets ejected from volcanoes, abraded and rounded clasts of igneous rock, liquid condensates from hot gases, splashes of melted rock from high-energy impacts, dust balls melted by lightning discharges, and clumps of solid material melted by rapid deceleration as they fell toward the sun. This is a distressingly long list of possibilities, but it must be remembered that chondrule formation is a process with which we have no experience. Chondrules do not occur in any other type of rock, except for a few isolated spheres in some lunar rocks formed by impact melting. This might suggest a similar origin for meteoritic chondrules; however, impacts by themselves do not appear to be capable of making rocks composed almost entirely of closely packed chondrules.

We now know some additional facts about these enigmatic miniature marbles. For instances, experiments that reproduce the internal textures of chondrules by cooling molten material at different rates have permitted the original cooling rates, generally less than several degrees per minute, to be determined. Additionally, relict grains in chondrules that have somehow been spared from complete melting have been recognized. A wealth of information about the chemical and isotopic compositions of chondrules has also been gathered. These kinds of data can be used to constrain theories of chondrule origin, but thus far no consensus has emerged. Chondrules are as much a puzzle to us now as they were to Sorby.

The components of chondrites that have been most frequently studied in recent years are the irregularly shaped white inclusions. Because of their distinctive chemistry, these are commonly called **calcium-aluminum inclusions**. These are especially abundant in certain types of carbonaceous chondrites, but they occur in ordinary chondrites as well. Calcium-aluminum inclusions take many forms, but most appear to have some kind of concentric structure formed by layers of different minerals. The minerals in these inclusions tend to crystallize at high temperatures, and for this reason they are thought to have been some of the first materials in chondrites to form. Some inclusions have internal textures that are similar to chondrules and signify that they crystallized from liquids, but others are ambiguous. Crystallization experiments on inclusions have reproduced their textures by starting from temperatures at less than complete melting and cooling fairly slowly at, say, one degree per minute. Some inclusions also contain tiny nuggets of platinum and other rare metal alloys, called **Fremdlinge** (German for "little strangers").

A FUZZY VIEW OF THE EARLY SOLAR SYSTEM

Driving in heavy fog is an uncomfortable experience for most people, because the fog limits our ability to see signs, landmarks, and other traffic. Some features in chondrites act in a similar way to obscure our ability to read the record of early solar system processes. Most chondritic meteorites have been affected to some degree by later events that have altered chondrites from their original state; 4.5 billion years is a long time for chondritic matter to remain untouched by other happenings, even in the relative isolation of space. We shall now examine to what extent chondrites have survived without change. Although we are primarily concerned with identifying chondrites in which the records are least altered, we shall see that the secondary events experienced by many chondrites are interesting in themselves.

A nuclear reactor requires a cooling system, because one of the by-products of fission is heat. We have already discussed the decay of radionuclides in chondrites, and some isotopes, especially those that decay rapidly, were probably capable of raising the temperatures of chondrites. Because rocks are notoriously poor conductors of heat, heat generated inside a chondrite parent body could not readily escape. This is thought by some meteorite researchers to be the cause of chondrite **metamorphism**, the adjustment of materials in meteorites to new thermal conditions. Another possible source of heat was electric current produced by fields induced by very high intensity solar winds. Observations of some very young stars, called T-Tauri stars, suggest that these eject large quantities of mass, and such strong solar winds may be characteristic of young stars. Others have suggested that chondrites have accreted from material at different temperatures, and those clumps that formed at higher temperatures would metamorphose themselves before cooling.

By what mechanism heating occurred is not known, but it is clear that many chondrites were thermally metamorphosed. During metamorphism, individual mineral grains recrystallized, consequently blurring the distinctive grainy texture of chondrules and inclusions embedded in finer-grained matrix. In severely heated chondrites, outlines of the original chondrules may be unrecognizable. Mineralogical changes also occurred during metamorphism. Volumetrically the most important minerals in chondrites are olivine and pyroxene, both silicates capable of wide ranges of magnesium-iron substitution. In unmetamorphosed chondrites, the magnesium and iron contents of these minerals are observed to

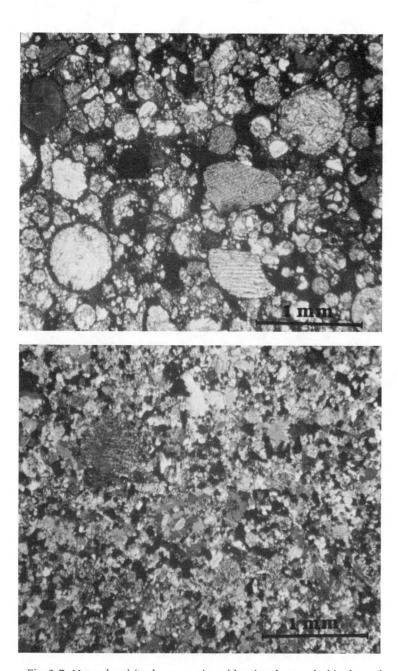

Fig. 2.7. Many chondrites have experienced heating that resulted in thermal metamorphism. The most obvious effect of this process is recrystallization, the formation of new mineral grains in the solid state. These new grains are more homogeneous in their chemical compositions than the corresponding grains in unmetamorphosed meteorites. These photographs show magnified views of an unmetamorphosed (type 3) ordinary chondrite from Tieschitz (Czechoslovakia) (top) and a metamorphosed (type 6) chondrite from Guarena (Spain) (bottom). Recrystallization at high temperature of the type 6 chondrite has blurred the distinctive chondrite texture.

vary widely, but metamorphism caused a narrowing of this range as individual grains attempted to equilibrate with each other at higher temperatures. Another change was the crystallization of glass to form feldspar grains in chondrules. All glasses are inherently unstable and tend to form crystals spontaneously; however, the process is slow and may require some time. Antique window glass may become clouded because of partial reorganization of the atoms into very small crystals. This process is accelerated at higher temperature, so that glasses are now absent from metamorphosed chondrites.

Six levels or grades of metamorphism in chondrites based on observable textural and mineralogical changes have been recognized. These metamorphic grades (called **petrologic types**) are combined with the chemical classes already discussed to formulate the currently used classification for chondrites. A portion of this classification is reproduced in Figure 2.8. Also shown are the approximate temperature intervals for each metamorphic grade, estimated from analyses of various minerals whose compositions are temperature-sensitive. The classification provides a convenient shorthand notation for chondrites, for example, E5 for an enstatite chondrite metamorphosed to petrologic type 5. Note in this figure that the least metamorphosed ordinary and enstatite chondrites start at type 3, whereas carbonaceous chondrites extend downward to type 1. This confusing situation occurs because the distinction between the temperature at which the meteorite accreted and the effects of later thermal metamorphism was not made. Some carbonaceous chondrites are less than type 3 because they accreted initially at lower temperatures than type 3 ordinary and enstatite chondrites. Individual boxes in the figure also contain the proportions of chondrites in each category. It is apparent that most ordinary and enstatite chondrites are metamorphosed, but most carbonaceous chondrites are not.

This is not to say, however, that carbonaceous chondrites are not altered in any way. Many of these meteorites seem to have suffered **aqueous alteration** by circulating fluids at low temperature. It has been estimated that alteration of one C2 chondrite required that a volume of water roughly equal to that of the rock at temperatures less than 20°C had been flushed through the meteorite. Alteration of C1 chondrites took place in a warmer, wetter environment, possibly 140°C and water-to-rock ratios of 3:4. The original matrix in most carbonaceous chondrites, a mixture of fine-grained olivine, pyroxene, and smaller amounts of other minerals,

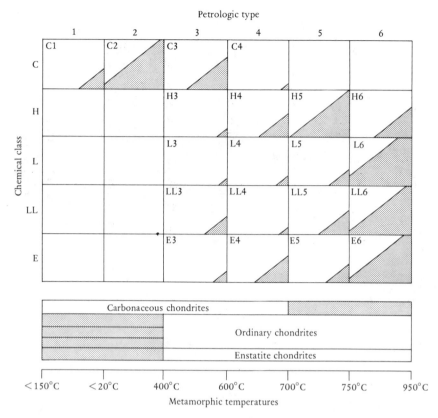

Fig. 2.8. *The commonly used classification system for chondrites consists of a matrix of chemical classes, identified by letters, and petrologic types, identified by numbers. Meteorites are divided into chemical classes according to their compositions and into petrologic types by their observable properties. Each chemical class presumably represents meteorites derived from one parent body, and the petrologic types are samples with different metamorphic histories. The shaded areas correspond to the relative percentages of petrologic types in each chemical class. The lower diagram shows that most ordinary chondrites are of high petrologic type, and chondrites of types 1 and 2 are unknown except for carbonaceous chondrites. Estimated temperatures required to produce the various petrologic types are shown across the bottom of the figure.*

has been transformed into an array of hydrous minerals called phyllosilicates. These are extremely complex, with layered structures somewhat similar to those of terrestrial clays. Fluids also moved through fractures in C1 chondrites and precipitated carbonate and sulfate minerals in veins. C3 chondrites are mostly unaffected by aqueous alteration, but C2 chondrites are altered, and C1 chondrites have been affected quite severely. Therefore, the classifica-

*Fig. 2.9. Aqueous fluids have permeated C1 and C2 chondrites and have al-
tered mineral grains into other phases. This microscopic view of the Orgueil
(France) C1 chondrite shows that fluids traveling in fractures deposited veins of
carbonate and sulfate minerals, which appear white against the dark matrix of
altered phyllosilicates.*

tion system for chondrites introduced earlier can be reinterpreted
to take aqueous alteration into account, as shown in Figure 2.10.
Considering the degree of alteration that C1 chondrites have experienced, it is surprising that their compositions still match that
of the sun so closely. This alteration must have occurred without
significant chemical exchange with the surroundings.

It is obviously of interest to know when these secondary events
took place. The rubidium-strontium isotopic clocks for chondrites
of high petrologic type have certainly been reset by the severe metamorphism that they experienced. Their ages (about 4.45 billion
years or older) suggest that metamorphism occurred within 100
million years of chondrite formation. Metamorphic heating must
have followed closely on the heels of chondrite parent-body accretion. This short time interval is consistent with the idea that such
rapid heating was caused by decay of short-lived radionuclides or
electromagnetic induction. Aqueous alteration was likewise an early
process. Calculations suggest that liquid water from ice melted by
decay of radionuclides could be retained for only a few hundred
million years. Evidence of now-extinct ^{129}I in magnetite, a pre-

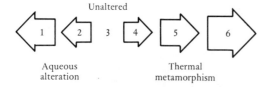

Fig. 2.10. *The recognition that aqueous alteration has affected type-1 and type-2 chondrites requires a reinterpretation of the original meaning of the chondrite classification scheme in Figure 2.8. In this new view, type-3 chondrites are nearly unaltered. Higher petrologic types were formed by thermal metamorphism, but lower petrologic types represent increasing degrees of aqueous alteration and thus have little or no temperature significance.*

sumed alteration product in carbonaceous chondrites, also suggests that aqueous alteration occurred very soon after accretion.

At some point in time, chondrites were excavated and liberated from their parent bodies. High-velocity impacts between orbiting meteorite parent bodies can eject crushed material from their surfaces or, in some cases, completely fragment such bodies. Collisions were probably common in the early solar system, so it seems possible that many meteorites may have experienced several impacts prior to being extracted from their parent bodies. Impact processes produce **breccias**, fragmental rocks composed of angular fragments of varying sizes and shapes. One study found that 25 percent of H, 10 percent of L, and 62 percent of LL chondrites were breccias containing rock **clasts**. In most cases, chondrite breccias contain no fragments of chondrite types other than their own (e.g., H, L, LL, C, E), but the clasts that are present may vary in metamorphic grade. This suggests that each parent body exhibited a range of metamorphic grades, and impacts presumably excavated the more deeply buried (highly metamorphosed) material and mixed it with the unmetamorphosed surface material. Brecciation among carbonaceous chondrites has not been reported as commonly as in ordinary chondrites, but it is more difficult to see and has probably been overlooked in many cases.

Another phenomenon that occurs in breccias is **shock metamorphism**. The collision of two incoming projectiles traveling at relative velocities of several kilometers per second instantaneously produces immense pressures. Values of at least 75 gigapascals (approximately 750,000 times the earth's surface atmospheric pressure) have been documented in some chondrites, and this may have been accompanied by significant heating. Some meteoritic

Fig. 2.11. Impacts into the surfaces of chondrite parent bodies have fragmented the target materials, and in many cases this rubble has been subsequently compacted into coherent rocks. One example is this chondrite breccia from Nakhom Pathom (Thailand). Clearly visible angular fragments are contained within a darker, pulverized matrix formed of finely comminuted chondritic material. Photograph courtesy of the Smithsonian Institution.

target materials have been partially melted, and others have experienced deformation of their crystal structures.

Such impact processes have affected chondrites throughout their history, but they were probably most potent in the early solar system before planets swept up most of the smaller orbiting debris. Most impacts appear to have taken place after thermal metamorphism, which occurred in the first 100 million years or so after chondrite formation, because chondritic breccias typically contain clasts of different metamorphic grades. However, some impacts probably occurred while metamorphism was still happening. Shock metamorphism can sometimes be dated by radioactive methods. The daughter isotopes of a few radionuclides are gases. One example is an isotope of gaseous argon, ^{40}Ar, which forms from the decay of ^{40}K, an isotope of potassium. Shock metamorphism can disturb a meteorite sufficiently for any argon gas present to escape. As the meteorite subsequently cools through a certain critical temperature, ^{40}Ar again begins to accumulate, and the radioactive clock is reset. The age derived from this kind of isotopic system, called a gas-retention age, represents the most recent thermal disturbance that affected the meteorite. For chondritic breccias this disturbance was shock metamorphism. Most measured chondrite gas-retention ages range from 4.4 to 0.5 billion years, reflecting random impact events.

All of these later events affecting chondrites provide an interesting picture of chondrite parent-body evolution since about 4.5 billion years ago. However, the record of early solar system processes in chondrites can be obscured to varying degrees by these overprints. It may seem surprising to non-geologists that we can read the primary record in chondrites through this haze of thermal metamorphism, aqueous alteration, brecciation, and shock effects. It is possible through careful observations to sort out these secondary events in meteorites and in some cases to compensate for their effects. However, most of what we know about early solar system processes comes from the chondrites that have suffered the least modification, the chondrites of petrologic type 3.

READING THE RECORD

In order to understand the significance of the bewildering components in unaltered type-3 chondrites, we must digress a bit to discuss current conceptions of how stars form. Astrophysical models indicate that a gravitationally contracting cloud of interstellar gas and dust will form a central mass concentration (which in the case of our solar system would ultimately become the sun), and rotation will flatten the outlying material into a disk, called the **solar nebula**. Previous calculations suggest that the inner parts of the nebula were sufficiently hot to vaporize all of the dust, and upon cooling this vapor would partly recondense as solids. Thermodynamic calculations have been employed to predict the order of appearance of minerals condensing from a cooling nebula of cosmic composition. This sequence in shown in Figure 2.12. Some minerals condense directly, but the **condensation** sequence is complicated by continuing reactions of already condensed minerals with nebular gases as temperatures are lowered further. The interesting thing about this model is that all of the minerals in this sequence occur in chondrites. The highest- temperature phases – corundum, perovskite, melilite, spinel, diopside – make up the bulk of calcium-aluminum inclusions. The minerals in the middle temperature range – olivine, pyroxene, metallic iron, plagioclase – occur as constituents of chondrules. The low end of the temperature spectrum is represented by minerals in matrix – iron-rich olivine and pyroxene, magnetite, troilite, and so forth. Do chondrites have cosmic compositions because they contain almost all of the nebular condensation sequence? This was the prevailing belief until just a few years ago.

More recent nebular models indicate that compression of gas and dust during contraction would not have produced temperatures high enough to vaporize interstellar grains. Thus, recondensation could not have occurred, except possibly for minerals that formed at low temperatures, that is, chondrite matrix materials. Other observations also support this view. The textures of chondrules and some calcium-aluminum inclusions clearly indicate that they crystallized from liquids. These may have been heated to their melting points in the nebula, but relict grains in chondrules suggest that they did not condense as liquids that subsequently crystallized. Chondrules and inclusions might be visualized as partly melted, residual nuggets left over when the more volatile elements were distilled off.

Isotopic studies of individual chondrules and calcium-aluminum inclusions provide another look at nebular processes. Oxygen has three stable isotopes, ^{16}O, ^{17}O, and ^{18}O. These can be separated from each other, or fractionated, by certain geologic events, but **fractionation** is always proportional to the differences in the masses of the isotopes. Thus, any process that increases ^{17}O relative to ^{16}O

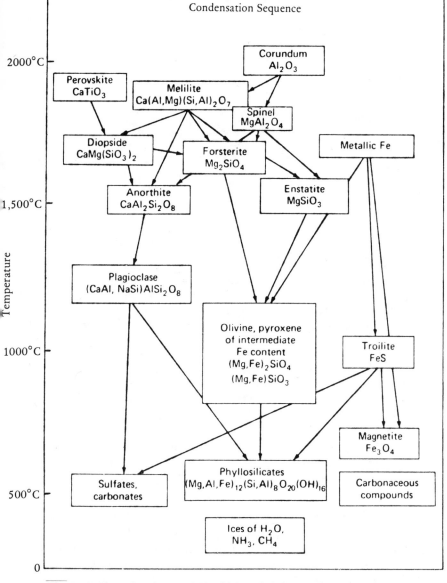

Fig. 2.12. The order of appearance of minerals from a cooling solar nebula has been predicted from theoretical calculations and is summarized in this figure. Minerals like perovskite, melilite, and corundum that form at high temperatures will condense directly from the gas, whereas those that form at lower temperatures will result from reactions of gas with previously condensed minerals (illustrated by arrows). Although such a condensation sequence is probably an oversimplified view of the formation of solid matter in the early solar system, it does predict the occurrence of the minerals that make up the bulk of chondritic meteorites.

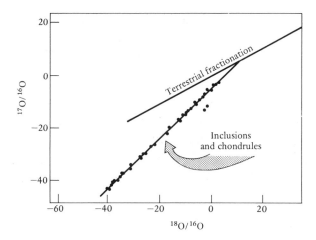

Fig. 2.13. *The isotopes of oxygen can be used as tracers to reconstruct processes that formed the components of chondritic meteorites. In this diagram the ratios of three isotopes of oxygen, relative to a standard material (average ocean water), are shown. All terrestrial samples fall along the line labeled "terrestrial fractionation," and any process that fractionates isotopes according to their masses will produce a similar distribution line with slope of $+\frac{1}{2}$. However, calcium-aluminum inclusions and chondrules in carbonaceous chondrites plot along a line with slope of $+1$. The most plausible explanation for this pattern is that these objects formed by the mixing of two isotopically distinct materials, one similar in composition to other solar system materials, and the other an exotic component rich in ^{16}O. This exotic component could have been introduced into the solar nebula when a nearby star exploded.*

by a small amount will increase ^{18}O relative to ^{17}O by that same amount, and ^{18}O relative to ^{16}O by twice that amount. If we plot the ratios of $^{17}O/^{16}O$ versus $^{18}O/^{16}O$, fractionated isotopes will be smeared along a straight line with slope of $+\frac{1}{2}$, as shown in Figure 2.13. All terrestrial materials fall along this line. If a sample starts on the line, there is no way for it to get off by natural geologic processes. However, the oxygen isotopic compositions of calcium-aluminum inclusions and chondrules in carbonaceous chondrites fall along a line with a different slope, as shown in the same figure. How could such isotopic variations arise? This second line extrapolates to pure ^{16}O at the origin of the figure, and thus compositions along the line could be produced by mixing pure ^{16}O with "normal" solar system oxygen, say at the intercept of this mixing line with the terrestrial fractionation line. The mystery then becomes, Where did the pure ^{16}O come from, and how was it mixed with normal solar system material? To produce pure ^{16}O, we require some kind of nuclear process, and the only reasonable possibility seems to be a **supernova**, the explosion of a massive star. During such an explosion, any ^{17}O or ^{18}O in the star theoretically would be destroyed by the intense nuclear reactions, and only ^{16}O would survive. Grains formed from supernova debris and containing pure ^{16}O could then wander into our solar nebula. When these were mixed in various proportions with other, normal solar system matter to make inclusions and chondrules, their bulk isotopic compositions would fall along the observed isotopic mixing line. These isotopic anomalies indicate that the solar nebula was not isotopically homogeneous, as would be appropriate for very hot nebula models, in which all solids would have been vaporized and thoroughly mixed.

Oxygen isotopes are not the only isotopic anomalies measured in calcium-aluminum inclusions. Excess amounts of an isotope of magnesium, ^{26}Mg, relative to terrestrial rocks have been found in inclusions. This isotope is the decay product of ^{26}Al, an aluminum radionuclide with an extremely short half-life, a mere 750,000 years. The ^{26}Mg occurs in mineral sites normally occupied by aluminum, indicating that it formed from ^{26}Al incorporated in the inclusion before it decayed. ^{26}Al, like the other short-lived radioactive isotopes we have already mentioned, is produced during a supernova. The rapid uptake of ^{26}Al in inclusions before it decayed suggests that the supernova that produced isotopic anomalies in chondrites must have occurred in our cosmic neighborhood.

This rather astounding finding initially met with some resistance from astronomers. However, it now appears that the proximity in space and time between this gargantuan explosion and the creation of our solar system was no accident. Astronomical observations of the remnants of material ejected by other supernovae reveal new stars in the making. The shock wave associated with an expanding supernova cloud compresses interstellar gas and dust ahead of it, forming clots of matter than may evolve into nebulae and stars. Thus, the supernova that apparently occurred in our part of the universe about 4.5 billion years ago may have triggered the formation of our solar system.

RAW MATERIALS FOR LIFE

An important chemical component of chondrites that has not yet been discussed is **organic matter.** These materials are complex, and sometimes very large, molecules composed mostly of carbon, hydrogen, oxygen, and nitrogen. Such materials probably occur in all chondrites but have been studied extensively only in carbonaceous chondrites. It is appropriate to mention contamination before we assess the results of organic analyses, as the quantities of some of these hydrocarbons are equal to the amounts that would be transferred to the meteorite by just a few fingerprints. Many chondrites have been in less than sterile conditions in museum collections for decades or may have been exposed to terrestrial weathering before collection. Therefore, analyses of organic matter in meteorites must be viewed with caution.

The discovery of small (1 millimeter in diameter or less) complex clumps of organic matter in chondrites spurred controversy some years ago. These so-called organized elements were thought by some to be biological in origin and were touted as evidence that primitive life forms were carried to the earth by meteorites. However, these were apparently contamination products, such as airborne pollen grains implanted in the meteorites while on the earth. At least one example in the Orgueil (France) meteorite appears to have been an elaborate hoax.

However, most of the organic material in chondrites does appear to be indigenous. This material is a very complex mixture of straight to slightly branching hydrocarbon chains (alkanes), rings (aromatic hydrocarbons), and carboxylic and amino acids. Organic materials formed by biological activity have the capacity to twist (circularly polarize) light passing through them in one direction and are said to be left- or right-handed. This ability is related to the fact that organisms synthesize compounds with certain preferred geometric forms. Nonbiological synthesis shows no such preference, and the organic fraction of chondrites consists of equal numbers of left- and right-handed molecules. Although it was not biological, the exact mechanism of origin of the organic material is not clear. There are two hypotheses, both suggested by chemist H. C. Urey. The first employs moderate heating of carbon monoxide, water, and ammonia in the presence of a catalyst. This is the Fisher-Tropsch synthesis, a commercially important reaction for the production of industrial organic compounds. The other contender involves a mixture of methane, water, and ammonia that is stimulated by a brief burst of energy such as lightning. A famous laboratory experiment by S. L. Miller, one of Urey's students, simulated this process, and it has come to be known as the Miller-Urey synthesis. Neither of the processes acting alone seems to be completely satisfactory.

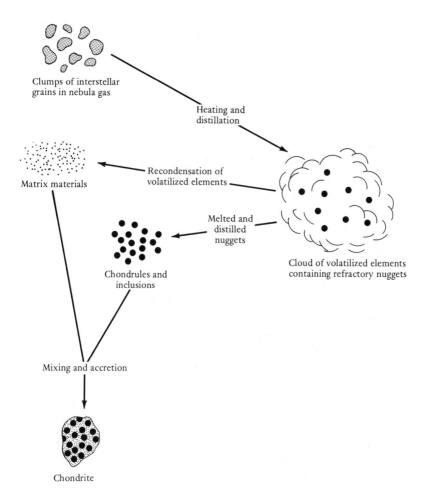

Fig. 2.14. The formation of chondrites probably occurred in a series of complex steps that are as yet poorly understood. One possible scenario is illustrated above: Interstellar grains were heated by some process in the nebula, resulting in melted droplets or refractory residues from which volatile elements had been driven off. During subsequent cooling, volatilized elements and other nebular gases condensed at moderate temperatures to form fine-grained solids. These two kinds of materials were then physically mixed and aggregated together to form chondrites.

The amino acids in chondrites are particularly important, as they are parts of the basic organic structures employed by living organisms. In fact, all of the organic bases in DNA, the fundamental carrier of hereditary information for life as we know it, have now been found in carbonaceous chondrites. Even though life forms were not carried to the earth in meteorites, the more basic constituents that made the evolution of life possible may have been.

THE MIRACLE OF CREATION

Eighteenth century scholars had had little success in deciphering hieroglyphs carved on old Egyptian monuments until French engineers found a tablet in 1799 while repairing a fort at the mouth of the Nile River near the town of Rosetta. On this tablet, the now-famous Rosetta Stone, are hieroglyphs and their translations in Coptic and Greek. This tablet unlocked the history of an entire ancient culture.

In a very real sense, chondritic meteorites are Rosetta Stones that have enabled us to decipher the earliest history of the solar system. The average chemical composition of the solar system and the timing of its birth are measurable only in chondrites. These rocks also provide a glimpse at the kinds of matter that accreted to form the planets. The diverse mineralogical components of chondrites provide an unparalleled record of early solar system processes, despite the overprints of later events that altered chondrites to varying degrees. Evidence for the nature of the collapsing solar nebula and for the nearby supernova that triggered its formation are found in these components. Organic compounds in these meteorites may have been the basic building blocks for the evolution of life. In a very real sense, chondrites permit us a glimpse into the miracle of creation.

SUGGESTED READINGS

Most of these readings are at a significantly more technical level than this book, because elementary descriptions of chondrites simply have not been written. However, they are all interesting contributions that will provide thorough treatments of chondrites for the serious reader.

GENERAL

Dodd R. T. (1981) *Meteorites: A Petrologic-Chemical Synthesis*, Cambridge University Press, 368 pp. (A superb technical treatise on the description and origin of meteorites; Chapters 2–6 describe chondrites in detail.)

King E. A., editor (1983) *Chondrules and Their Origins*, Lunar and Planetary Institute, Houston, 377 pp. (A collection of 25 technical papers summarizing recent research on chondrules.)

CHONDRITE CLASSIFICATION

Van Schmus W. R. and Wood J. A. (1967) A chemical-petrologic classification for the chondritic meteorites. *Geochimica et Cosmochimica Acta* 31, 747–765. (A classic technical paper that first outlined the presently used chondrite classification scheme.)

Wasson J. T. (1974) *Meteorites*, Springer-Verlag, Berlin, 316 pp. (A complete descriptive reference volume for meteorites, including extensive classification tables for chondrites and other meteorites.)

METAMORPHISM AND ALTERATION

Dodd R. T. (1969) Metamorphism of the ordinary chondrites: A review. *Geochimica et Cosmochimica Acta* 33, 161–203. (Technical review describing thermal effects in chondrites.)

McSween H. Y. Jr. (1979) Are carbonaceous chondrites primitive or processed? A review. *Reviews of Geophysics and Space Physics* 17, 1059–1078. (Technical review describing aqueous alteration in chondrites.)

CHONDRITIC BRECCIAS

Wilkening L. L. (1977) Meteorites in meteorites: Evidence for mixing among the asteroids. In *Comets, Asteroids, and Meteorites*, edited by A. H. Delsemme, University of Toledo Press, pp. 389–396. (Technical paper describing chondrite breccias.)

AGES

Papanastassiou D. A. and Wasserburg G. J. (1969) Initial strontium isotopic abundances and the resolution of small time differences in the formation of planetary objects. *Earth and Planetary Science Letters* 5, 361–376. (Technical paper presenting radiometric ages for chondrites.)

CHEMICAL COMPOSITION

Anders E., Hayatsu R., and Studier M. H. (1973) Organic compounds in meteorites. *Science* 182, 781–791. (Technical paper reviewing research on the organic constituents of chondrites.)

Grossman L. and Larimer J. W. (1974) Early chemical history of the solar system. *Reviews of Geophysics and Space Physics* 12, 71–101. (Technical review of the inorganic constituents of chondrites.)

Wasson J. T. (1985) *Meteorites, Their Record of Early Solar System History,* W. H. Freeman, New York, 267 pp. (Appendix D is a tabulation of the cosmic abundances of the elements.)

THE EARLY SOLAR SYSTEM

Schramm D. N. and Clayton R. N. (1978) Did a supernova trigger the formation of the solar system? *Scientific American* 239 (4), 124–132. (Nontechnical paper describing the evidence for a supernova just prior to the formation of the solar system.)

Wood J. A. (1979) *The Solar System,* Prentice-Hall, Englewood Cliffs, N. J., 196 pp. (An excellent introductory text; Chapters 6 and 7 describe nucleosynthesis in stars and the formation of the solar nebula.)

3 Chondrite parent bodies

The primitive nature of chondrites demands that they come from objects that have somehow escaped geologic processing. This is most easily accomplished if the parent bodies are not large planets. In this chapter we shall explore the possibility that chondrites are derived from the smaller objects in the solar system, **asteroids** and **comets**. The major observational distinction between these two kinds of bodies is that comets are luminous objects and asteroids are not. After the initial discovery of asteroids, there was a flurry of astronomical activity to catalog these "minor planets." However, by the 1950s, interest had waned to such a degree that major observatories considered it improper to study such "vermin of the sky." This denigration of status occurred because until just a few years ago asteroids were considered merely superfluous material that was unable even to accumulate into a modest-sized planet. However, new observational techniques have rekindled interest in asteroids. A recent comment by one astronomer indicates that the study of these bodies has now gained some measure of legitimacy:

... while asteroids may well be characterized as the garbage of the solar system, there is nothing undignified or trivial about delving into their secrets.

Comets are more spectacular as astronomical objects, but their small sizes probably also qualify them as cosmic junk.

METEORITE ORBITS

In 1959, an H5 ordinary chondrite impacted at Pribram, near Prague, Czechoslovakia. What made this fall unusual was that it was accidentally photographed by several cameras originally set up to track artificial satellites. Because accurate **orbits** for meteorites can be ascertained only from precise observations made simultaneously at two or more locations, Pribram became the first recovered chondrite whose orbit prior to earth capture could be determined.

Partly as a consequence of the inadvertent Pribram experiment,

Fig. 3.1. This fireball was photographed by the Prairie Network station at Hominy, Oklahoma, in early 1970. The picture has been tilted so that the ground is at a 45° angle in the lower right corner. Spaces between luminous segments of the fireball's trajectory are caused by a chopping shutter used for timing and velocity determination. The faint lines in the background are star

in 1964 a network of cameras was set up in the United States by the Smithsonian Astrophysical Observatory to photograph the trajectories of any meteoroids of sufficient size to reach the ground intact. This "Prairie Network" consisted of 16 stations disposed over a circular area approximately 500 kilometers in radius and centered on southeastern Nebraska. The system functioned for about ten years. Although the orbits of hundreds of fireballs were recorded during the network's lifetime, only one meteorite with a recorded orbit was eventually found. A similar network in Canada recovered an additional meteorite. Both of these, Lost City (Oklahoma) and Innisfree (Alberta), were ordinary chondrites. This was somewhat disappointing, but not surprising, considering the frequency of ordinary chondrite falls. The orbits of several other ordinary chondrites, Farmington (Kansas) and Dhajala (India), have been calculated from independent observations of the falls by eyewitnesses. Their calculated trajectories are not as accurate as those determined from photographic networks, but they do improve this rather limited data base.

The orbits for all five chondrites are shown in Figure 3.2. Each of these paths is highly elliptical and crosses inside the earth's orbit, as it must if the chondrite is to fall on the earth. But the really interesting feature of each of these orbits is that its **aphelion**, indicated on the figure by a dot, lies between the orbits of Mars and Jupiter. This is the position of the **asteroid belt**, primarily located between 2.2 and 4.0 AU (an AU is an **astronomical unit**, the mean distance between the sun and the earth).

In this belt there are about 2,500 asteroids of sufficient size that their orbits are well known, and many more thousands that are less than 1 kilometer or so in diameter. The largest asteroid is **1 Ceres**, with a diameter of 1,025 kilometers. Despite the large number of asteroids and the fact that 45 of them have diameters greater than 200 kilometers, all of them taken together have only a fraction of the mass of the moon. For most people, the term "asteroid belt" conjures up visions of fields of closely spaced chunks of rock stretching as far as the eye can see, probably because of the navigational hazard they have posed in so many science-fiction

trails resulting from the earth's rotation during the three-hour exposure. These photographic results were used to pinpoint the location of this meteorite, which was recovered a few days later near Lost City, Oklahoma. Its pre-atmospheric orbit was calculated from sightings by a number of network stations. Courtesy of the Smithsonian Astrophysical Observatory.

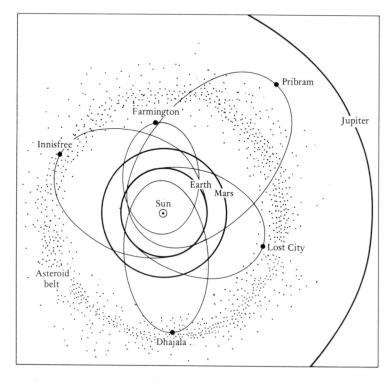

Fig. 3.2. The calculated orbits of recovered meteorites provide information on the sources of these objects. Five recovered ordinary chondrites had highly ellipt- ical orbits. All of these had aphelia, the approximate locations of which are illustrated by small dots in this figure, in or near the asteroid belt between Mars and Jupiter. This suggests that chondrites may be fragments of asteroids. The orbits are drawn to scale, but their orientations are chosen for clarity of illustra- tion.

movies. In reality, on their way to Jupiter, a number of NASA spacecraft have traversed the asteroid belt "blind" without a scratch. Some years ago it was argued that all of the asteroids were once assembled into one now-disrupted planet. However, calculations indicate that the swarm of asteroids could never have accreted into a larger body because of the perturbing effects of the planet Jupi- ter. The massive gravitational field of this giant neighbor would have ripped apart any larger planet within its sphere of influence as quickly as it formed.

The fact that the aphelia of all five meteorite orbits lie in this particular region suggests that chondrites may be derived from as- teroids. Such a link was first postulated in 1805 by the German astronomer H. Olbers, but of course without the supporting evidence of meteorite orbital calculations.

Not all asteroids are confined entirely to orbits within the main belt; some have more elliptical trajectories that cross into the inner solar system, much like the orbits of the recovered chondrites. The earliest discovered examples of such near-earth asteroids were given the names of the more active and sometimes erotic figures of Greek mythology, such as Apollo, Amor, Eros, and Adonis. The number of asteroids has by now outstripped the available list of mythical names, and new asteroids, whether near-earth or main-belt, are named after whomever or whatever the discoverer wishes. The name is normally preceded by a number corresponding to the order of discovery, as in 887 Alinda or 1915 Quetzalcoatl. Many near-earth asteroids cross the path of the earth and may properly be considered very large meteoroids.

It would, of course, be interesting to have a close-up view of asteroids, particularly if some are the parent bodies of chondrites. Their surface appearances, sizes, and shapes would provide useful information on their evolutionary history, but this kind of information is difficult to obtain. "Asteroid" actually means "starlike," and viewed through a telescope, these planetesimals are merely point sources of light. Even the largest known asteroid has a disk less than a second of arc across, which means that it is similar in appearance to many stars. A novel experiment was performed to assess the dimensions of the near-earth asteroid 433 Eros during a close approach in 1975. According to calculations completed just two hours prior to the event, Eros was to eclipse a star in the constellation Gemini. Teams of observers rushed to preselected locations in Connecticut and recorded the duration of the occultation at various positions along the path of the moving shadow. These observations made it possible to calculate the size of this asteroid as approximately 7 by 19 by 30 kilometers. This irregular slab tumbles end over end every 5.3 hours.

The two moons of Mars are also irregular in shape and are no larger than modest-sized asteroids: Diemos measures 10 by 12 by 16 kilometers, and Phobos 20 by 23 by 28 kilometers. Both objects have highly unusual orbits that suggest they may have been asteroids that were gravitationally captured from the nearby main belt. The tiny moons were photographed by Mariner 9, and two portraits of Phobos are reproduced in Figure 3.3. Phobos has a heavily cratered surface, including one crater with a diameter that is one-fifth of that of the body. The crater Stickney, large enough to justify

Fig. 3.3. The two tiny moons of Mars are likely to be captured asteroids. Although we have never had a close look at an asteroid, these Viking images of Phobos from different orientations probably are similar to asteroids. The large impact feature at the bottom of the photograph on the right is crater Stickney. The fracture system on the other side of Phobos, shown on the left, suggests that the satellite was nearly disrupted by the impact that produced this large crater. Violent collisional histories are probably common among asteroids. Photographs courtesy of T. C. Duxbury (Jet Propulsion Laboratory).

having its own name, is associated with radial grooves that may represent fractures. These features testify to a massive impact that may have nearly fragmented the entire planetesimal. The orbit of Phobos is dangerously close to Mars. Because it is still being accelerated and drawn closer, Martian tidal forces will reduce it to an orbiting ring of small fragments within the next 50 million years.

The view that asteroids are chondrite parent bodies agrees with our previous assessment of chondrites as more or less unprocessed, early solar system materials. By preventing planetesimals from assembling into a larger planet capable of geologic processes, Jupiter may have preserved chondrites in their present, relatively pristine states. However, the irregular shapes and cratered surfaces of the few possible asteroids that have been examined at close range suggest that such planetesimals may have experienced repeated collisions.

ANOTHER WAY TO LOOK AT ASTEROIDS

Dogs and bats can hear sounds beyond the accessible frequency range of human hearing, and many birds of prey have incredible visual capacities. Man has learned to stretch his capabilities by harnessing parts of the electromagetic spectrum outside the range of his own limited perceptions, and one notable application has been in the study of asteroids. A panoply of new observational techniques available to astronomers has transformed asteroids from starlike objects into individual little worlds. Because asteroids generate no light of their own, all of these methods depend on the properties of sunlight reflected from asteroidal surfaces. The ratio of light reflected by a surface to the incident light is its **albedo**, a measure of the efficiency of the reflection process.

The first technique that we shall consider is **spectrophotometry**. In a previous discussion of isotopes and radiometric dating, we considered only the protons and neutrons in the nuclei of atoms. An atom can be envisioned as a small pea (the nucleus) suspended at the center of a mass of cotton candy, the latter representing the distribution of electrons about the nucleus. The movement of electrons within this density cloud provides the basis for spectrophotometry. Individual electrons follow certain paths, within regions called "orbitals," but electrons can jump to outer vacant orbitals by absorbing extra energy of appropriate wavelengths. However, the energy necessary depends on the geometry of the surrounding atoms. Analysis of the energy absorbed by electrons moving between orbitals can provide information on the identity of an element and its coordination to other atoms. Sunlight provides a continuum of wavelengths from which the electrons may select. The spectrum of sunlight reflected from the surface of a mineral will be missing those wavelengths that are absorbed by electrons, the exact wavelength being characteristic for certain atoms in specific kinds of minerals. The energy ingested by crystals may have wavelengths in the visible, infrared, or ultraviolet ranges. Light of longer wavelength than visible light is infrared; that of shorter wavelength is ultraviolet. For example, the most abundant element that exhibits this behavior is iron, which absorbs energy in the visible and near-infrared parts of the spectrum when it is situated in crys-

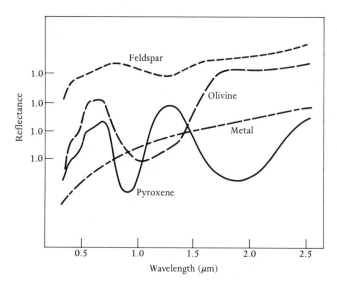

Fig. 3.4. The spectrum of sunlight reflected by certain minerals shows absorption bands (valleys) at certain wavelengths because of ingestion of energy by electrons as they jump between orbitals. This forms the basis for a method of meteorite identification by remote sensing. Shown here are examples of the reflectance spectra for feldspar, olivine, pyroxene, and iron-nickel metal. Meteoritic material consisting of a mixture of these minerals would have a composite spectrum formed by the integration of these individual curves.

tallographic sites within minerals. This is the cause of visible color in many iron-bearing minerals.

Examples of the **reflectance spectra** for some individual minerals are illustrated in Figure 3.4. If these minerals are combined to form a rock, the resulting spectrum is a messy composite of the individual spectra for the constituent minerals. However, these overlapping **absorption bands** provide a kind of signature for asteroidal surfaces. Examples of reflectance spectra for asteroids are shown in Figure 3.5. The curves in the lower part of this diagram are for dark objects with low albedos, and curves in the upper reaches of the diagram are for brighter objects with high albedos. Each of the spectral curves has a peculiar shape, generally dominated by absorption bands from very few minerals. This is not to imply that just a few minerals are present, but rather that only a few minerals produce important absorption bands. As a general rule, a linear spectrum indicates a dominant iron-nickel metal component, and a curved spectrum with absorption bands indicates silicates like pyroxene or olivine.

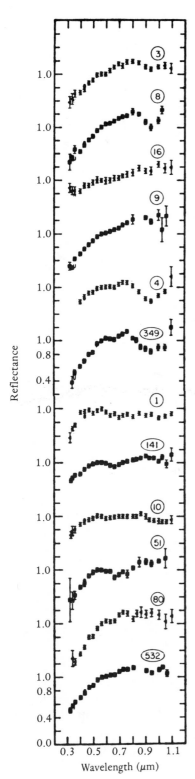

Fig. 3.5. Illustrated here are the reflectance spectra for 12 asteroids. Each of these curves is a composite of overlapping spectra for different minerals. The circled numbers refer to catalog numbers of asteroids: 3 Juno, 8 Flora, 16 Psyche, 9 Metis, 4 Vesta, 349 Dembowska, 1 Ceres, 141 Lumen, 10 Jygiea, 51 Nemansr, 80 Sappho, and 532 Herculina. Included in this list are representatives of most of the different spectral types of asteroids.

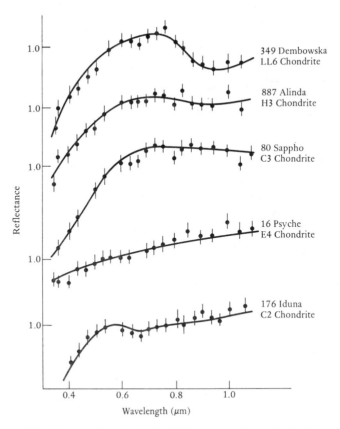

Fig. 3.6. In order to try to identify the compositions of asteroids, comparisons are made between asteroidal and meteorite reflectance spectra. This diagram illustrates five such comparisons between selected asteroids (dots with analytical error bars) and certain classes of chondrites (curves). Because absolute albedo depends on particle sizes, which are unknown in the case of asteroids, it is permissible to translate asteroidal spectra up or down in the diagram in order to obtain a match. All data are therefore arbitrarily assigned an albedo value of 1.0 at a wavelength of 0.56 micrometer.

The spectral reflectivities of various types of powdered meteorites have also been measured in the laboratory. Spectral curves for some typical chondrites are illustrated in Figure 3.6. These measurements provide an interesting basis for comparison with the asteroidal data. Because absolute albedo depends not only on mineralogy but also on particle sizes and packing, individual curves for meteorites or for asteroids can be translated up or down in these diagrams. As a consequence, the relative shapes of spectral curves are more important for comparisons than the absolute albedo levels. If we adjust albedo levels at some arbitrary wave-

Table 3.1. *Some asteroidal compositional types*

Type	Albedo (%)	Spectral reflectivity	Possible meteorite analog
E	High (>23)	Featureless	Enstatite chondrite
R	High (>23)	Very red	Enstatite chondrite
S	Moderate (7–23)	Red, absorption band at 0.9 to 1.0 micrometer	Ordinary chondrite
C	Low (2–7)	Neutral, slight blue absorption	Carbonaceous chondrite
D	Low (2–7)	Very red	Carbonaceous chondrite

length to a common value, quite remarkable matches between asteroidal and chondritic spectra can be obtained. A few examples are illustrated in Figure 3.6.

Albedo and reflectance spectra provide the basis for a classification system that groups similar asteroids (Table 3.1). Type-C asteroids have silicate and carbon mineralogies, indicated by relatively flat spectra with low albedos. These dark planetesimals appear to be possible sources for carbonaceous chondrites. 1 Ceres, the largest asteroid, is a somewhat unusual C type, but its spectral properties provide a convincing link with carbonaceous chondrites. The spectrum of water-bearing clay minerals on its surface has also been recognized; these clays are reminiscent of the mineralogy of the matrix of C1 and C2 chondrites. Ceres also apparently has water frost or ice on its surface. Type-D asteroids also have very low albedos, possibly due to organic material, and may be carbonaceous as well. Type-S asteroids have higher albedos and more pronounced absorption bands, indicating the presence of olivine, pyroxene, and iron-nickel metal. At least some of these are possible candidates from which ordinary chondrites might be derived; however, there are some other kinds of meteorites composed of silicates and metal for which S asteroids may also be parent bodies. Regrettably, E chondrites do not have diagnostic spectra in the visible and infrared regions. They are distinctive in the far ultraviolet, but the earth's atmosphere blocks these wavelengths. Type-E and Type-R asteroids have generally featureless spectra and very high albedos, consistent with enstatite chondrites, which contain abundant metal. There are also several additional asteroid types that we have not mentioned, to which we shall return in later chapters.

Spectra of near-earth asteroids show that they differ in composition from most of the asteroids in the main belt; however, a few of these could have chondritic compositions. For example, 2100 Ra-Shalom has a spectrum similar to those of carbonaceous chondrites, and 1862 Apollo is like ordinary chondrites.

Another technique employed in the remote sensing of asteroids is **radiometry**, a type of heat sensing. Thermal vibrations are in the infrared region of the spectrum, and infrared measurements are used to detect heat escaping from poorly insulated buildings, or in military applications to spot moving vehicles in complete darkness. A variation of this method also permits us to estimate the sizes of asteroidal bodies, if some assumptions can be made about the composition of asteroidal surfaces (from spectral reflectivity). Comparison of the thermal brightness with the brightness at visual wavelengths provides a means of calculating asteroidal diameters to an accuracy of about 10 percent.

Polarimetry is the study of how light is polarized during reflection from an asteroid. Polarized light vibrates in one preferred direction. The lenses of polarizing sunglasses restrict the passage of light except for that vibrating in only one direction, and thereby greatly diminish the light intensity reaching the eye. In a somewhat similar manner, fine particles on an asteroidal surface may produce reflected light that is vibrating in a preferred direction. This property provides information on surface texture and roughness, and when combined with other data it serves as an independent check on albedo and diameter measurements.

Speckle interferometry is a relatively new technique that yields crude but resolvable images of the larger asteroids. Data on several asteroids suggest that they may have co-orbiting satellites, much like massive boulders rotating as a single system. This and other developments, when perfected, will no doubt tell us a great deal more about asteroids.

STRUCTURE OF THE ASTEROID BELT

Asteroidal spectra in general are more varied than the spectra of known meteorite types. This reinforces the idea that meteorites have provided only a limited and probably biased sample of what is out there. Moreover, the relative abundances of asteroidal types are not anything at all like those of fallen chondrites. Type-C asteroids far outnumber all other types, possibly composing three fourths of the main belt, in contrast to the relative scarcity of car-

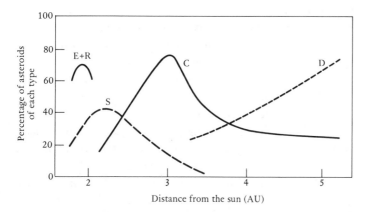

Fig. 3.7. Synthesis of all the available reflectance spectra suggests that the proportions of different asteroid types vary systematically with distance from the sun. E and R types, which are possible enstatite chondrite parent bodies, are located in the inner fringe of the asteroid belt. Ordinary chondrites may be derived from S-type asteroids that tend to occur slightly farther out. C and D types occur primarily in the outer belt and may be the sources of carbonaceous chondrites.

bonaceous chondrites. Type-S asteroids – the possible parent bodies for ordinary chondrites – make up only about 17 percent of the belt. Only the extreme, metal-poor end members of the S group have spectra that approach those of ordinary chondrites, so it is possible that the parent bodies for the most common type of chondrite falling to earth are rare or totally lacking in the main belt.

The asteroidal classes are also not distributed uniformly within the belt. The inner fringe of the asteroid belt is where most E and R asteroids are located, succeeded outward by S, C, and D, as illustrated in Figure 3.7. This pattern accords well with that predicted from the properties of chondrites. Enstatite chondrites are more highly reduced than ordinary chondrites and should have been formed at higher temperatures in the nebula, that is, closer to the sun. Relative to ordinary chondrites, carbonaceous chondrites have higher volatile-element abundances and are more highly oxidized, as is appropriate for meteorites that formed in cooler regions of the nebula located farther from the sun. This distribution pattern may suggest that many asteroids are still located in the orbital positions in which they formed. An alternative view is that the inner main belt is a dumping ground in which a significant quantity of asteroids originally formed in the inner solar system were later implanted.

Within the asteroid belt, there are objects of many different sizes. Astronomers have noticed a curious distinction between the size distributions for C- and S-type asteroids. When plotted on a logarithmic diagram of diameter versus abundance, type-C asteroids form a straight line. Such a linear relation is known to be characteristic of a population of objects whose sizes have been determined by fragmentation due to mutual collisions. When type-S asteroids are plotted on the same diagram, they exhibit a nonlinear trend, with a "hump," diagnostic of incomplete fragmentation. We therefore expect carbonaceous chondrite parent bodies to be more thoroughly disrupted by repeated impacts. This is a logical outcome of the fact that carbonaceous chondrites have lower crushing strengths than ordinary chondrites.

If many asteroids have been fragmented by collisions, it seems natural to assume that they will have irregular shapes. We have already seen that the near-earth asteroid 433 Eros and the two tiny moons of Mars are irregular in shape. The tumbling motion of an elongated asteroid in space can be detected from measured variations in brightness with time. It appears from such measurements that many asteroids may be nonspherical. Shape irregularity cannot be maintained for C-type asteroids larger than about 100 kilometers in diameter, because the force of gravity will collapse these relatively weak objects into spheres. However, the greater strength of large S-type asteroids permits their irregular shapes to persist against higher gravity.

In summary, the relative proportions of asteroid types in the main belt appear to differ from the abundances of various chondrite types. Most asteroids appear to be fragments of larger bodies disrupted by impacts, but a few large bodies, such as 1 Ceres, have thus far escaped the collisional fate of most asteroids. Despite mixing of asteroid orbits within the belt, some regularity in the spacing of compositional types of asteroids relative to distance from the sun occurs.

SAMPLING PLANETESIMALS

What we can learn about the parent bodies of chondrites is not strictly limited to astronomical observations. The chondrites themselves provide a record of processes that occurred on or within these bodies, and from these processes we can infer additional facts about these planetesimals that cannot be gained from remote observations of their surfaces. In 1964, a chemist, E. Anders, drawing almost exclusively from meteorite data, was the first scientist to formulate a detailed model for chondrite parent bodies. He argued in favor of asteroids, a clear departure from the moon- or planet-sized bodies advocated by previous workers.

One important question that may be addressed from studies of chondrites is how many kinds of meteorites formed on a given parent body, or, put another way, how many parent bodies are necessary to explain the observed chondrite population? In our previous discussion of oxygen isotopes in chondrites, we learned that isotopic fractionation always produces a line of slope $+\frac{1}{2}$ on a plot of $^{17}O/^{18}O$ versus $^{16}O/^{18}O$, but mixtures of different isotopic compositions are not constrained to such a line. If different chondrite parent bodies incorporated varying amounts of two mixed components, they could theoretically plot anywhere along the line joining the end members. Any subsequent fractionation processes would then smear the isotopic compositions of individual samples along lines with slope $+\frac{1}{2}$, parallel to the terrestrial line. Each chondrite parent body could have its own oxygen isotopic composition, although it is possible that several bodies might have the same isotopic signature. Therefore, the number of distinct points or lines on an oxygen isotope diagram represents a minimum required number of parent bodies. Figure 3.8 shows oxygen isotopic analyses for chondrites. This is only a tiny portion of the diagram we have seen before, because the isotopic differences in whole chondrites are much less exaggerated than those in calcium-aluminum inclusions. H chondrites are distinct from L and LL chondrites, necessitating at least two parent bodies for ordinary chondrites. Carbonaceous chondrites also require several planetesimals. The enstatite chondrites are likewise distinct from ordinary and carbonaceous chondrites and interestingly plot on the terrestrial line. Following this reasoning, it appears that enstatite chondrites are most like the material that accreted to form the earth. Oxygen isotopic variations suggest a minimum of five different chondrite parent bodies. However, other differences in chemistry between L and LL chondrites seem to require separate parent objects for these, and one or two additional planetesimals are probably required for altered carbonaceous chondrites whose primary isotopic compositions have been modified by exchange with water.

We have already noted that chondrite breccias almost always contain fragments of the same chemical class. If two chondrite classes, say H and E, were formed on the same parent body, they would likely have been mixed into breccias by repeated impacts.

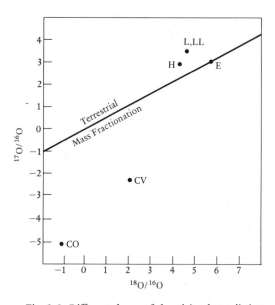

*Fig. 3.8. Different classes of chondrites have distinct oxygen isotopic composi-
tions, indicating that they must have been derived from different parent bodies.
Despite the fact that L and LL chondrites are similar in isotopic composition,
other chemical differences indicate that these are also from distinct planetesimals. Enstatite (E) chondrites plot on the terrestrial mass-fractionation line,
suggesting that of the various chondrite classes, these are most like the materials
that accreted to form the earth. CO and CV refer to two subgroups of C3 carbonaceous chondrites. Isotopic compositions of C2 and C1 chondrites have been altered during aqueous alteration and are not shown. All isotopic ratios shown
here are relative to a standard, average ocean water.*

The scarcity of foreign clasts also suggests that collisions between
asteroids of the same composition must have been much more
common than collisions between asteroids of different classes. The
radial distribution of asteroidal types observed from spectral reflectivity provides an explanation for this, in that impacts are most
likely between objects traveling in similar orbits. The properties of
chondritic breccias therefore also support the idea of distinct parent bodies for chondrites of different classes.

ASTEROID HEATING

Fragments in chondritic breccias, although of the same class, commonly differ in terms of metamorphic grade. This observation can
be interpreted in two ways: Either colliding planetesimals were of

the same class, but had been heated to different degrees, or similar impacting bodies contained both metamorphosed and unmetamorphosed materials. The second idea seems much more plausible. Rocks are notoriously poor conductors of heat, and as a consequence heat was unlikely to have been distributed evenly throughout the interior of even small planetesimals. This would have produced zones of differing metamorphic grades arranged radially within each body.

Before we examine some thermal models for chondrite parent bodies, let us briefly review some facts we have learned about metamorphism from studies of ordinary chondrites. This information will provide some basic constraints on modeling. Peak temperatures in chondrite parent bodies ranged from about 750–950°C for type 6 to 400–600°C for unmetamorphosed type-3 material. If the relative abundances of chondrite types have any significance, more highly metamorphosed chondrites predominate over unmetamorphosed ones. The rubidium-strontium isotopic age data for type-3 and type-6 chondrites indicate that the time interval between parent body-formation and the end of metamorphism was on the order of only 100 million years. If we accept this interval as the time required for metamorphosed chondrites to cool to 400°C, the approximate blocking temperature for the movement of rubidium and strontium atoms, we can estimate a cooling rate of a few degrees per million years. This value can be combined with measurements of the rate at which heat flows through chondrites to estimate the size of ordinary chondrite parent bodies. The value for the diameter calculated in this way is about 170 kilometers, a very reasonable size for asteroids. The most likely heat sources were rapid decay of short-lived radionuclides or electric currents induced by the early solar wind (**electromagnetic induction**). Notice that the first heat source is internal and the second is external to the body; consequently, an internally heated planetesimal would have higher metamorphic temperatures in the inside, in contrast to the pattern produced by external heating.

Detailed thermal models of the parent bodies of H and L chondrites have been formulated using the foregoing constraints and the assumption of internal heating by short-lived radionuclides. A schematic model for the H chondrite body from such calculations is shown in Figure 3.9. The parent object for the L chondrites is similar to this, but because of compositional differences it attains slightly higher internal temperatures and consequently produces a

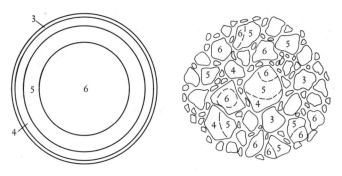

Fig. 3.9. *These sketches compare two models for the internal structures of chondrite parent bodies. On the left is a diagram of an onion-shell model for the H chondrite parent body. Heating by rapid decay of the short-lived isotope* ^{26}Al *produced a metamorphosed (petrologic type 6) interior, but slow conduction of heat allowed the surface layers to remain relatively cool (type 3). The diagram on the right shows a brecciated rubble pile formed by accretion of smaller, already metamorphosed planetesimals. A similar asteroid could also form by gravitational reassembly of a larger onion-shell body that had been disrupted by a large impact.*

greater proportion of type-6 material. Hypothetical parent bodies like these with concentric zones of different metamorphic grades are called **onion-shell models**. One of the nice results of these models is that they reproduce the various petrologic types of chondrites in similar proportions to chondrite falls.

If metamorphism occurred in internally heated bodies, some of the highly volatile elements that could be thermally mobilized should have migrated to the cooler regions near the surfaces of the bodies, where they may have recondensed. Very high contents (above cosmic abundances) of such mobile elements have been measured in some type-3 ordinary chondrite breccias that presumably formed on or near the surfaces of their parent bodies. This suggests that even unmetamorphosed chondrites may have felt the effects of thermal metamorphism happening deeper in the interiors of their parent objects.

It also seems possible that heating by short-lived radionuclides might occur in objects smaller than the 170–kilometer-diameter body considered previously. The heat produced from radioactive isotopes like ^{26}Al can be estimated, but we know almost nothing about the rate of accretion. Metamorphism of an asteroid of several hundred kilometers diameter necessitates accretion times of several hundred thousand years or less. If accretion were an order of magnitude slower, taking several million years, maximum temperatures would be reached earlier in smaller bodies with diameters of just a few kilometers.

External heating by induced electric currents would probably have a very short duration because of limitations on T-Tauri activity of the sun. Because the heating interval would be short relative to the time required for heat to conduct into the interior, a planetesimal experiencing metamorphism driven by this process would have to be small. Diameters of only 4 to 20 kilometers have been estimated for externally heated planetesimals. As noted earlier, such bodies would have reversed zoning, with type-6 chondrites on the outside.

Is there any other means by which we can constrain the sizes of metamorphosed chondrite parent bodies? Pressure increases with depth in a planetesimal because of the weight of the overlying rock. Pressure at any point within the body is a function of rock density and gravitational acceleration, in turn related to the size of the body. For chondrite parent bodies with diameters of a few hundred kilometers, the most extreme pressure experienced is less than 0.15 gigapascal (equal to 1.5 kilobars, approximately 1,500 times the earth's surface atmospheric pressure). There are numerous minerals or assemblages of minerals whose compositions are pressure-sensitive, but unfortunately these are applicable only at higher pressures. Substitution of extra sodium and aluminum in chondritic pyroxenes is dependent on pressure, and small increases in these elements occur in chondritic pyroxenes with increasing metamorphic grade. The total amount of this component suggests pressures of less than 1 kilobar. Such data are consistent with an internally heated body, but the low pressures occurring in asteroid-size objects have thus far precluded an accurate assessment of their sizes during metamorphism.

Postmetamorphic cooling rates are also related to size. In principle, material deep in the interior of a body is effectively insulated and should cool more slowly than that at or near the surface. There are several ways in which cooling rates can actually be measured; in effect, these are cooling speedometers. The techniques of using the **metallographic cooling rate** depends on the diffusion of nickel from one iron-nickel metal alloy (taenite) to the other (kamacite) at high temperatures. Nickel moving to the edges of taenite grains forms a compositional gradient that can be measured with an electron microprobe.* Movement of nickel atoms ceases below about 500°C, so the nickel compositional gradient is frozen in at this temperature. From measured profiles of nickel concentrations in taenite, it is possible to calculate the rate at which these grains cooled through 500°C.

Another method for determining cooling rates employs the fission tracks that were earlier noted as evidence of extinct radionuclides. Although these tracks are produced continuously, minerals at high temperatures anneal themselves and erase these tiny bullet holes. However, different minerals begin to retain tracks at different temperatures. For example, the track retention temperature for pyroxene is about 580°C, and that for whitlockite is about 380°C. If the tracks were produced by decay of ^{244}Pu, then from the density of tracks we can estimate the time interval required for the meteorite to have cooled from 580°C to 380°C.

These two completely independent ways to estimate cooling rates generally give comparable results. Measured cooling rates for individual ordinary chondrites range from 1 to 1,000°C per million years. Such cooling rates correspond to burial depths ranging from only a few kilometers for the fast rates to as many as 100 kilometers for the slowest. This would seem to argue that metamorphism occurred within larger parent bodies. There is one problem with this interpretation: cooling rates should be related to petrologic type. The cooling rates for the most severely metamorphosed samples should be slower than for lower petrologic types if internal heating occurred, or vice versa for external heating models. However, there is no observed correlation between cooling rate and metamorphic grade, as is illustrated in Figure 3.10. For this reason, it has been proposed that peak metamorphic temperatures were realized within small bodies with diameters of only a few kilometers. In this view, metamorphism occurred during rather than after accretion. Assembly of these still warm bodies into **rubble piles** of several hundred kilometers diameter resulted in chaotic mixing of all metamorphic grades at different depths. Cooling rates were

---

* The *electron microprobe* is capable of performing chemical analyses on very small spots within individual mineral grains of a rock sample. A beam of high-energy electrons is focused onto a tiny spot on the polished surface of the rock. The sample is viewed through a microscope, and any area of interest can be moved under the beam. The electrons excite elements in the mineral to emit x-radiation. The wavelengths of x-rays are characteristic for different elements, and their intensities are functions of element concentrations. Crystal spectrometers record the x-rays emitted, and a computer program converts this measurement into the concentration of each element within the analyzed spot.

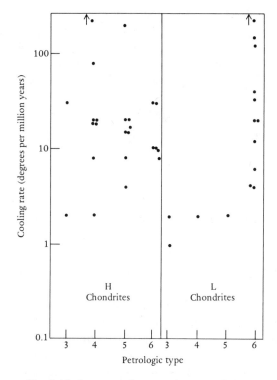

Fig. 3.10. Because rocks are such poor conductors of heat, material in the deep interior of an asteroid should cool more slowly than that near the surface. Therefore, onion-shell models like that shown in Figure 3.9 predict an inverse correlation between petrologic type and cooling rate. Studying the diffusion of nickel through metal grains in chondrites provides a means of determining the rate of cooling. Illustrated in this diagram are the metallographic cooling rates as a function of petrologic type for two classes of ordinary chondrites. Highly metamorphosed chondrites of types 5 and 6 clearly did not cool more slowly than their less metamorphosed relatives of types 3 and 4. This observation is a problem for onion-shell models for chondrite parent bodies, but it may be explained by rubble-pile models.

then controlled by depth of burial in the larger reassembled planetesimals. Such a parent body is illustrated in Figure 3.9. In support of this model, it has been noted that fragments in some chondrite breccias had to have cooled at drastically different depths. It is not possible to excavate material from many tens of kilometers depth on a planetesimal of 100 to 200 kilometers diameter by impact without breaking up the body, and theoretical studies of the collisional history of asteroids suggest that disrupted bodies of sufficient size would tend to reassemble because of gravity effects.

Despite the absence of a correlation between petrologic types and cooling rates, it may yet be possible to salvage onion-shell models. An insulating layer of dust accreted on the outside of a large body may have permitted slower cooling rates than would otherwise have occurred. Alternatively, reaccretion (if rapid enough) of still-hot metamorphic chunks of a disrupted onion-shell body could have modified the subsequent cooling rates. At present it does not seem prudent to rule out either onion-shell or rubble-pile models for heated asteroids.

CHONDRITIC DIRT

Scientists formerly visualized the lunar surface as marred by sharp fractures and angular, rocky prominences. The first high-resolution photographs from the moon, however, indicated rather subdued, rolling topography. The reason for this miscalculation is a mantling layer of soil, called the **regolith** (Greek for "rocky layer"). Regolith is defined as a pervasive blanket of loose, rocky material that rests on coherent bedrock. The Apollo astronauts noted that their rocket exhausts created flurries of rock dust, often as fine as flour. However, soil samples collected on the moon also contain blocks of various sizes. Regoliths form because of repeated bombardment by small meteroids that pulverize the surface rocks, occasionally punctuated by larger impacts that excavate blocks from below and mix them with dust.

This soil-forming process is naturally not restricted to the moon. Viking photographs of the Martian surface indicate development of a regolith, and such features are probably characteristic of all planets with cratered surfaces. The asteroid-size moons of Mars have regoliths, and it seems reasonable to assume that asteroids do as well. Polarization of light reflected from asteroids has been interpreted as indicating dust-covered surfaces of unknown thickness.

At present, our best source of information about asteroidal regoliths comes from chondrites, provided, of course, that asteroids are chondrite parent bodies. Even if unconsolidated regolith dust could reach the earth, it could not survive atmospheric transit. Fortunately, well-indurated soil clumps, called **regolith breccias**, sometimes form because of compaction and cementation of the various soil ingredients. Such regolith breccias compose about half of the carbonaceous chondrites and smaller percentages (generally about 10 percent or less) of ordinary and enstatite chondrites.

Fig. 3.11. This footprint on the lunar surface was made by the boot of astro naut Neil Armstrong. The sharpness of the indentation shows that the soil consists of a fine, cohesive powder. The regolith also contains blocks of rock and was formed during continued impacts by meteoroids of all sizes. This kind of pulverized surface layer is probably characteristic of chondrite parent bodies. Photograph courtesy of NASA.

However, not all chondritic breccias were formed in regoliths. Recall that asteroids may be rubble piles, in effect, giant breccias reassembled from smashed asteroids, and most of these materials would be buried below the zone of active regolith formation. Only the outer veneer of breccias would be reworked into regolith.

The textures of all breccias, regolith and otherwise, are similar and consist of angular rock and mineral fragments of varying sizes. However, true regolith breccias display some distinctive characteristics that allow us to recognize their mode of formation. On bodies without atmospheres, surficial material is subjected to bombardment by energetic nuclear particles. This irradiation has left unmistakable fingerprints in some chondritic breccias. Significant amounts of gases with the distinctive isotopic composition of the **solar wind** were implanted on the surfaces of individual grains. Microcraters and particle tracks also formed as more energetic ions penetrated into grains. An array of new isotopes was further produced by the disruption of incoming particles, and these can be measured. Because cosmic rays and solar wind provide particles with different energies, the irradiation effects of the more energetic

Fig. 3.12. Although they have weak gravitational fields, small solar system bodies like asteroids can apparently retain enough comminuted impact ejecta to form regoliths. That this is possible is demonstrated by a close-up view of the surface of Phobos, one of the tiny moons of Mars photographed by Viking. A regolith blankets the landscape and produces a subdued appearance despite the presence of craters. Photograph courtesy of T. C. Duxbury (Jet Propulsion Laboratory).

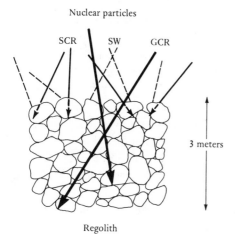

Nuclear particles

SCR   SW   GCR

3 meters

Regolith

Fig. 3.13. The surface regolith on an asteroid is irradiated by an assortment of tiny particles, as illustrated in this sketch. The highest-energy particles are galactic cosmic rays (GCR), which penetrate 2 to 3 meters and produce particle tracks and new isotopes as they strike appropriate target atoms. Less energetic solar cosmic rays (SCR) form particle tracks to depths of only about 1 millimeter. Solar wind (SW) particles are so weak that they are deposited only on the outermost parts of grains lying on the surface. Study of regolith irradiation provides a means of determining the age of a regolith. Most of these particles are shielded from the earth's surface by its atmosphere.

particles extended to greater depths in regoliths, as illustrated in Figure 3.13. The record of exposure to particles of different energies can even be converted into an **exposure age** for each type of radiation. Typical times for surface exposure of chondrite regolith materials are thousands of years.

One interesting feature of regoliths is that their characteristics alter, or "mature," as their surface residence times increase. Exposure ages of lunar regolith samples are generally 100 to 10,000 times longer than for chondrite breccias. The greater age of lunar regolith breccias affords an instructive comparison with chondritic regolith samples. The most visible difference in maturity between these is the presence in the lunar regolith of glass and partially melted rock and soil fragments, called **agglutinates**. These melted products are the result of continued impacts, which increase with exposure time. Glass and agglutinates make up as much as 50 percent of lunar breccias, but are exceedingly rare in chondrites. The lunar regolith is also finer-grained than its chondritic counterparts.

Why should the surface residence time and maturity of regolith materials be so much less on asteroids than on the moon? One possible explanation hinges on the fact that the amount of orbiting material and the resulting impact rates are much greater in the asteroid belt than in the vicinity of the moon at 1 AU; consequently, more impacts might result in higher rates of regolith stirring and overturn. Asteroids have weaker gravity fields than the moon because they are less massive, and their short regolith residence times have also been attributed to easier ejection of regolith material into space. Mathematical models confirm that impact rate and parent body-size are the controlling factors in regolith formation, but point to a different and rather surprising conclusion. These calculations suggest that asteroidal regoliths are much thicker (on the order of a kilometer for 100- to 300-kilometer-diameter bodies) than the lunar regolith (measured at about 5 to 10 meters at the Apollo landing sites). The greater thickness results in part from the fact that the higher lunar gravity would produce less widespread ejecta. The same impact onto an asteroidal body would deposit ejecta over the entire body because of its lower gravity field. Thus, the explanation for the shorter exposure time for asteroidal regoliths may be that surface materials are buried faster under more rapidly evolving surfaces.

COSMIC SNOWBALLS

Passages of comets through the inner solar system are among the most spectacular astronomical events that can be seen with the naked eye. But such isolated cometary occurrences are but the tip of the iceberg, so to speak. Most comets belong to a vast swarm, called the **Oort cloud**, after the Dutch astronomer who proposed its existence, that surrounds the solar system. This comet reservoir defines a spherical volume extending out to as far as 50,000 AU; in contrast, the other bodies of the solar system all lie inside 40 AU and within a flat plane. The objects in the Oort cloud probably number in the billions. Current models suggest that comets may originally have formed in the vicinity of Jupiter, Saturn, Uranus, and Neptune, but they were scattered to the outer fringes of the solar system by gravitational encounters with these planets.

Once in the Oort cloud, comets drift around the sun until gravitationally influenced by the nearer stars. Such perturbations may occasionally send a comet into the inner solar system. These are the long-period comets, so named because of the long intervals between successive arrivals. Once inside 40 AU, a small fraction of

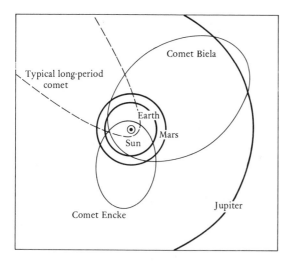

Fig. 3.14. Although the conventional cometary model is a dirty snowball of ice and dust, comets have been suggested as possible parent bodies for some kinds of chondritic meteorites. The aphelia of long-period comets are so far away that their orbits appear to be parabolic rather than elliptical. However, some long-period comets may change into short-period comets if their orbits are perturbed by interaction with planets. After a number of close passes to the sun, short-period comets may lose their ices, and only stony residues will remain. The observed orbits of several short-period comets like Encke and Biela are similar to those of near-earth asteroids, and thus these comets may evolve into earth-crossing asteroids of rocky material with time.

these may interact with planets and evolve into short-period comets that make moderately frequent passes into the inner solar system. The highly modified orbits of these are very similar to those of near-earth asteroids, as illustrated in Figure 3.14.

Comets take their name from the Greek word "kome" ("hair"), a reference to the spectacular, streaming tails that always point away from the sun because of interaction with the solar wind. Study of the tails provides some information on the composition of comets. Type-I tails are straight and are composed primarily of ionized gases. The spectra of such gases suggest that water ices may be the dominant constituent of cometary nuclei. Diffuse, gently curved type-II tails apparently consist of dust, indicating that some fraction of rocky material exists in comets as well. Individual comets may exhibit one or both types of tails.

The most widely accepted conception of a comet nucleus is that it resembles a **dirty snowball**. These frozen chunks of water, methane, and ammonia ices contain dust particles – silicates and possibly metals and oxides that were incorporated when the comets formed. The snowball model has sometimes led to the misconception that cometary nuclei are more ice than rock, when in fact the emitted material in tails contains as much dirt as snow. The diameter of the nucleus may measure only several kilometers, though the coma (the incandescent halo surrounding the nucleus) may measure 1,000 kilometers across.

The properties of comets are observed to change perceptibly with continued exposure to the sun, and the dirty-snowball model may offer a way to explain this. Long-period comets, which may be newly arrived from the Oort cloud, expel large quantities of dust and gas. After a number of trips around the sun, short-period comets appear to "burn out" as much of the icy material capable of being volatilized is expended. Comets have also been noted to fragment or disintegrate on close approach to the sun or after multiple passes.

The orbits of short-period comets are very similar to those of near-earth asteroids, leading to the suggestion that these asteroids represent the devolatilized nuclei of spent comets. One unusual near-earth asteroid, 2201 Oljato, may provide a more persuasive link with comets. The reflectance spectrum of this asteroid is highly unusual, and its features might be better interpreted as emission bands rather than absorption of energy. The implication of this is that the spectrum may result from the light-scattering effects of dust particles being ejected from the surface. The activity is too slight to make the asteroid glow like a comet, but it is suggestive of a prior cometary history.

The orbital similarities of short-period comets and near-earth asteroids seem to allow the possibility that chondrites may be pieces of comet nuclei. Given the icy composition of comets, however, it is improbable that they could be the parent bodies of thermally metamorphosed chondrites. If any meteorites are derived from comets, they are probably the C1 or C2 chondrites. Repeated close approaches to the sun may have melted ice, causing the relatively low temperature aqueous alteration processes that characterize these meteorites. However, this, too, seems unlikely. Any liquid water on comets due to solar heating would be promptly lost to space, and the amounts of regolith implanted solar-wind gases and the high impact rates experienced by carbonaceous chondrites point to the asteroid belt.

This is not to say, however, that meteoritic material of cometary

Fig. 3.15. The 1908 Tunguska explosion in a remote part of Siberia repre-sents a possible collision between a comet and the earth. Although the event left no crater, trees located as far as 10 to 15 kilometers away from ground zero were blown down radially away from the point of impact. This photograph of the destruction was taken by a Russian expedition to the site 20 years later. Photograph courtesy of the Smithsonian Institution.

origin never arrives on earth. In 1908, a massive explosion oc-curred in the sparsely populated Tunguska River region of Siberia, and its seismic disturbance was observed as far away as several thousand kilometers. From a vantage point 200 kilometers away, the object causing the explosion was described as "an irregularly-shaped, brilliantly white, somewhat elongated mass." Because of World War I and the Russian Revolution, exploration of the area by scientific teams was unfortunately delayed for many years. An expedition led by Soviet scientist L. A. Kulik finally reached the site in 1927 and found a scene of total devastation. The explosion left no large crater, but was apparently equivalent to a 10- to 15-megaton bomb detonated in the atmosphere. All trees were flat-tened radially away from "ground zero" for several kilometers. No meteorite fragments were ever recovered, although small glassy spherules were found in soil at the site in 1961. These facts are consistent with the hypothesis that the Tunguska event represented the collision of a comet with the earth.

The dusty trails left by comets produce meteor showers whenever the earth's orbit takes it through points where comets have passed. Interplanetary dust particles (micrometeorites) collected by high-altitude aircraft are likely candidates for such comet debris. These fragile aggregates of microscopic grains have mineralogical and chemical compositions that are broadly similar to those of carbonaceous chondrites, but there are enough differences to suggest that these particles are unique.

JUNK OR TREASURE?

Archeologists must reconstruct ancient settlements and infer something about past life-styles, often from painfully little evidence. Many of their best clues are provided by what once was worthless garbage cast aside by the populations under scrutiny. A few chips of pottery, a broken kitchen utensil or arrowhead, or some bones or seeds can provide valuable information on trade, industry, or diet.

In a somewhat analogous manner, we have been forced to reconstruct or infer the salient features of chondrite parent bodies from some limited astronomical observations and the accumulated record of post-accretional processes in the meteorites themselves. Neither of these methods of inquiry alone provides a comprehensive view of where chondrites originate, so an interdisciplinary approach is necessary. We have learned from orbital measurements that asteroids are probably the sources of chondrites, and the spectral properties of individual asteroids match those of the various kinds of meteorites. The parent bodies of distinct chondrite classes apparently formed under different nebular conditions prevailing at various distances from the sun. Thermal processes resulted in metamorphism of planetesimal interiors, while regolith development modified their surfaces. Fragmentation by collision with other asteroids and possible reassembly may have modified the basic structural framework of asteroids. Interplanetary dust may represent meteoritic materials derived from comets.

We began this discussion of asteroids and comets by noting that they were essentially debris left over from the formation of the more impressive planets of the solar system. The insignificant sizes

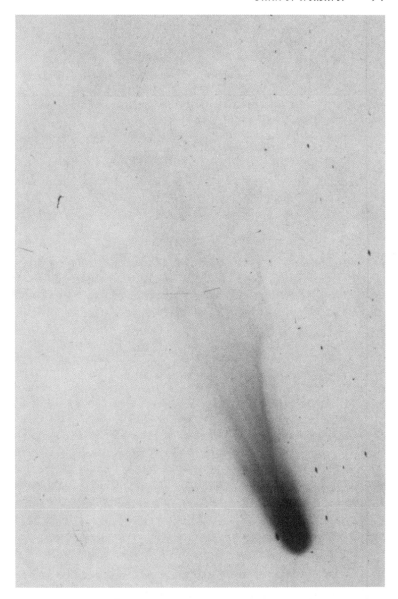

Fig. 3.16. This negative photograph shows the dusty trail left by Comet Halley as it passed through the inner solar system in 1910. This may be the source of some of the interplanetary dust particles collected in the earth's atmosphere.

of these bodies belie their importance in solar system history. Chondrite planetesimals provided the building blocks for other worlds like our own, and their natures and compositions controlled many of the most fundamental properties of planets.

SUGGESTED READINGS

Unfortunately, there are almost no nontechnical publications about asteroids. The first reference is very interesting without sacrificing detail. The other papers are demanding but excellent sources, especially for the reader with some astronomy background.

GENERAL

Hartmann W. K. (1983) *Moons and Planets*, 2nd edition, Wadsworth Publishing, Belmont, CA, 509 pp. (An excellent undergraduate textbook with chapters on meteorites, asteroids, and comets; this book is highly recommended for the nonspecialist.)

ASTEROID PROPERTIES AND ORBITS

Chapman C. R. (1975) The nature of asteroids. *Scientific American* 232 (1), 24–33. (A somewhat dated but informative nontechnical paper on asteroids.)

Chapman C. R., Williams J., and Hartman W. K. (1978) The asteroids. *Annual Reviews of Astronomy and Astrophysics* 16, 33–75. (Technical review of asteroid research.)

Wilkening L. L. (1979) The asteroids: Accretion, differentiation, fragmentation, and irradiation. In *Asteroids*, edited by T. Gehrels, University of Arizona Press, Tucson, pp. 61–74. (Technical paper describing the origin and evolution of asteroids.)

COMETS

Whipple F. L. (1978) Comets. In *Cosmic Dust*, edited by J. A. McDonnell, Wiley, New York, pp. 1–72. (Moderately technical paper summarizing comet observations and properties.)

REFLECTANCE SPECTRA

Gaffey M. J. and McCord T. B. (1978) Asteroid surface materials: Mineralogical characterizations from reflectance spectra. *Space Science Reviews* 21, 555–580. (Technical review of the application of reflectance spectra in determining asteroid compositions.)

THERMAL HISTORY AND MODELS

Miyamoto M., Fujii N., and Takeda H. (1981) Ordinary chondrite parent body: An internal heating model. In *Proceedings of the 12th Lunar and Planetary Science Conference*, pp. 1145–1152. (Technical paper that discusses the onion-shell model for asteroids.)

Scott E. R. D. and Rajan R. S. (1981) Metallic minerals, thermal histories and parent bodies of some xenolithic, ordinary chondrite meteorites. *Geochimica et Cosmochimica Acta* 45, 53–67. (Technical paper in which the rubble pile model for asteroids was first presented.)

REGOLITHS

Housen K. R. and Wilkening L. L. (1982) Regoliths on small bodies in the solar system. *Annual Reviews of Earth and Planetary Sciences* 10, 355–376. (Technical review of the formation of regoliths on asteroidal surfaces.)

CHONDRITE SOURCES

Anders E. (1975) Do stony meteorites come from comets? *Icarus* 24, 363–371. (Technical paper discussing possible sources for chondritic meteorites.)

Wasson J. T. and Wetherill G. W. (1979) Dynamical, chemical and isotopic evidence regarding the formation locations of asteroids and meteorites. In *Asteroids*, edited by T. Gehrels, University of Arizona Press, Tucson, pp. 926–974. (Technical paper summarizing the various lines of evidence for formation locations of small bodies.)

Fig. 4.1. Eucrites are the most common type of achondrite. This slab of the ALHA 76005 (Antarctica) eucrite has a brecciated texture produced by impacts into the crust of its parent body. Clasts of igneous rocks, like the one in the middle left of the slab, indicate that the components of this eucrite breccia originally formed as lava flows. The cube, measuring 1 centimeter on a side, indicates the scale. Photograph courtesy of NASA.

4 Achondrites

In early 1980, Mount St. Helens in the Pacific Northwest stirred from a century-long slumber and exploded with a force hundreds of times greater than that of the atomic bombs dropped on Hiroshima and Nagasaki in World War II. Such violent eruptions of ash and gas are among the most awesome spectacles that nature has to offer. Although they conform to the popular conception of volcanic events, these are not the kinds of volcanic eruptions by which most planetary surfaces are fashioned. Less obvious, but volumetrically more important, are the relatively quiet effusions of lavas like those that formed the ocean floors and large portions of some continents on the earth. These also have produced massive shield volcanoes, such as Mauna Loa in Hawaii, a mountain resting on the seafloor, as tall as Everest but of much greater bulk. These molten lavas solidify into **basalt**, a rock composed primarily of the minerals pyroxene and plagioclase. Eruptions of basaltic lavas appear to provide a common thread in the histories of planetary bodies. Such lava flows have sculpted the face of the moon and are well represented in rocks returned by the Apollo and Luna missions. Basaltic volcanic features have also been recognized on the surfaces of Mercury, Mars, and Venus, the latter viewed only indirectly by radar mapping through its dense atmospheric cloud cover.

The solidification of erupted molten material, called **magma**, produces igneous rocks. Volcanic products are only the surface expressions of vast underground igneous plumbing systems by which magmas rise to the surface. Magma that congeals before its ascent to the surface is complete produces **plutonic** rocks. Some meteorite parent bodies have also experienced igneous activity, and both **volcanic** and plutonic rocks are represented in meteorite collections. Meteorites of igneous origin are called achondrites.

The igneous nature of achondrites was first recognized by the German petrologist G. Rose in 1825. The name means "without chondrules," emphasizing the distinction between achondrites and chondrites, the other major class of stony meteorites. In this chapter we shall restrict ourselves to consideration of just a few of the known kinds of achondrites. These will be lumped into four associations that probably were formed through related igneous processes on four different parent bodies.

The study of achondrites (as well as terrestrial igneous rocks) is much more important that it might first appear, because igneous processes are the primary means by which planetary bodies evolve. Planetary evolution, like biological evolution, generally results in increasing complexity with time. Igneous processes permit a planet to differentiate, that is, to transform from an initially homogeneous body to one with compositionally distinct core, mantle, and crust. They also periodically resurface parts of planets and release volatile elements to form atmospheres, but the bulk of this evolution happens deep in planetary interiors. Igneous rocks may carry records of the compositions of their deep source regions, the nature and distribution of heat sources, the timing of melting events, the solidification process, and the global features that permitted magma generation and facilitated its ascent. Put more succinctly, igneous rocks provide windows through which we can observe evolutionary processes in planetary interiors. These geologic windows are often partly shuttered, so that we can only glimpse these otherwise inaccessible regions, but some samples provide such interesting views that we can be excused for gawking.

ORIGIN AND EVOLUTION OF MAGMA

The knowledge of igneous processes gained from several centuries of geologic studies of terrestrial rocks provides such an important foundation for understanding achondrites that it is prudent to review the fundamentals of igneous petrology before proceeding further. The diversity among terrestrial igneous rocks is much greater than that among achondrites, so we shall restrict this discussion only to basaltic rocks that are similar in composition to achondrites.

Basaltic magmas can exist as liquids only at temperatures of 1,000°C or more. How are such high-temperature melts generated? The heat necessary to melt rocks in the earth's interior ultimately derives from decay of radioactive elements. Temperatures in the deep interior increase because heat escapes to the surface very slowly. Measured temperature gradients are commonly on the order of 20 to 30°C for each kilometer of depth in the earth's crust. However, temperature is not the sole factor that determines whether or not rocks will melt. Pressure also increases with depth, as the mass of overlying material is gravitationally pulled toward the center of the earth. Increasing the confining pressure generally results in raising the melting point of rocks. It is this pressure increase that keeps the earth and other planets from having completely molten interiors. It then follows that deeply buried rocks may melt for one of two reasons. First, rocks melt if the temperature is increased to their melting point at the prevailing pressure; this could occur in response to some localized heat source. Alternatively, if temperature remains fixed, rocks may spontaneously melt when the pressure on them is lowered. This second case is a little harder to visualize, but such decompression may occur as plastically deforming but still solid rock is slowly pushed upward.

Unlike any pure substance such as ice, which melts at 0°C, rocks consist of mixtures of minerals that melt over a range of temperature, generally a few hundred degrees. As a rock is gradually heated or the pressure on it is slowly lowered, the low-melting-point fraction of the rock will begin to liquify. The first-formed melt will probably be different in composition from the original rock, although the magma composition will steadily approach that of the bulk source rock as melting progresses. However, experience indicates that partial melting produces most magmas, and complete melting is rarely if ever achieved. Despite the fact that magmas form by melting only a small fraction of source-region materials, it is sometimes possible to reconstruct their source rocks. Properties of the source region are imprinted on magmas derived from it, just as the characteristics of certain grapes are passed on to the wines made from them. As in the case of the wine connoisseur, the trick is to develop the skills to make these properties diagnostic.

After a certain quantity of magma has been generated, it may segregate from the solid residue and begin to work its way toward the surface. The driving force behind this upward movement is the fact that magmas are generally less dense than the overlying rocks. Dissolved gases such as water vapor may be held in solution at high pressure, but can form bubbles as the magma approaches the surface. This process, similar to the production of effervescent carbon dioxide as a can of soda pop is opened and depressurized, may accelerate ascent.

Upon cooling, magmas undergo **crystallization** to form igneous rocks. The motions of rapidly vibrating atoms in the melt decrease with lowering temperature until they are finally locked into crystals with orderly internal arrangements. Rather than crystallize all at once, a magma will develop numerous crystal nuclei that will grow until they impinge on one another. The liquid is eventually transformed into a mass of tightly interlocked crystals. The crystallization process is influenced by the rate of cooling, which largely controls the size of crystals. Atoms in rapidly cooled magmas quickly lose their motion and combine to form many crystal nuclei. The resulting texture consists of many small crystals, typical of volcanic rocks. Plutonic magmas cool more slowly, permitting atoms to migrate over large distances to the sites of growth and form large crystals.

Crystallization under plutonic conditions is complicated by the fact that different minerals form at varying temperatures during slow cooling. Early formed minerals that are more dense than the surrounding liquid may sink to form layers on the bottom of the magma chamber. Rocks formed from accumulated crystals are called **cumulates**. Other processes, such as moving currents within the magma body, may produce the same result, purging the melt of its crystals. The physical segregation of crystals from magma is called **fractional crystallization**, or **fractionation** in abbreviated form. This process changes the composition of the original melt, called the primary magma, by depleting it in the early crystallizing components. Fractionation, a very common process, can happen at any point en route to the surface. It is important to recognize its effects, because inferences about source regions drawn from such altered melt compositions may be erroneous.

A GEOCHEMISTRY LESSON

The igneous processes we have just described are summarized in Figure 4.2. Each of these events can impart its own telltale geochemical signature on magmas and igneous rocks. Before we can decipher the records of these processes in achondrites, we must digress briefly and discuss some rudiments of geochemistry.

We have already learned that some minerals can have complex compositions, such as olivine, which consists of a mixture of Mg_2SiO_4 and Fe_2SiO_4 end members. What has not been mentioned before is that the composition of olivine is dependent on the temperature at which it crystallizes. Olivine that forms at high temperature is much more magnesium-rich than that which is sta-

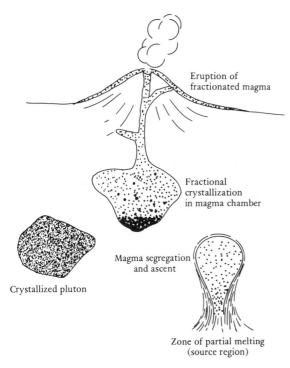

Eruption of
fractionated magma

Fractional
crystallization
in magma chamber

Magma segregation
and ascent

Crystallized pluton

Zone of partial melting
(source region)

*Fig. 4.2. This schematic drawing summarizes some of the processes that pro-
duce igneous rocks. Partial melting of a deep source region produces magma
that segregates from the unmelted residue and ascends to higher levels. Some
magmas crystallize relatively slowly at depth to form plutons. Most, but not all,
undergo some type of fractional crystallization in magma chambers that alters
their compositions. Some of these fractionated magmas may ultimately erupt on
the surface as lava flows that solidify rapidly into volcanic rocks.*

ble at low temperature. Pyroxenes behave similarly in terms of
their magnesium and iron contents. In the plagioclase series, the
calcium-rich end member is the high temperature component, and
sodium-rich plagioclase forms at low temperature. A consequence
of this temperature dependence is that accumulations of early formed
crystals, called cumulates, of olivine and pyroxene from a cooling
magma should be magnesium-rich, and plagioclase cumulates
should be calcium-rich. The complementary melt is of course driven
to higher iron and sodium contents by loss of these high-temperature minerals. This affords one means of recognizing these products of fractional crystallization. Turning this argument around,
we can see that a rock undergoing the agonizing experience of

partial melting will release its low-melting-point fraction first. Compared with the solid that is left behind, this melted fraction should be enriched in iron and sodium.

Elements that occur only in trace quantities can also be useful in understanding fractional crystallization and partial melting. Accurate concentrations of trace elements are difficult to measure and require sophisticated analytical techniques such as neutron activation.[*] The **rare earths** are a group of 15 elements from lanthanum (La, atomic number 57) to lutetium (Lu, atomic number 71) in the periodic table. We shall focus on these elements as examples of the utility of trace elements, but others may be equally informative. **Ions** of the rare earth elements are generally trivalent ($+3$ charge) and are fairly large in size. During crystallization, trace elements must find homes for themselves within ordinary minerals by substituting for other more abundant elements. However, because of their size and charge, almost none of the rare earths fit comfortably into most crystallographic sites. The one exception to this rule is europium (Eu, atomic number 63), which can form ions with $+2$ charge that are about the same size as calcium ions (Ca^{2+}). Europium is thus tailored to sneak into calcium sites in minerals like calcic plagioclase. This can be seen in Figure 4.3, a plot of the experimentally determined ratios of the concentrations of rare earths in various minerals relative to the liquid with which they coexist. Rare earths are excluded from all these minerals, except europium, which is partitioned into plagioclase, creating the so-called europium anomaly.

During fractional crystallization, rare earth elements are excluded from most minerals, but their concentrations build up in the remaining magma. As a consequence, cumulates should have low abundances and fractionated melts should have high abundances of these elements. If fractional crystallization involves plagioclase, the plagioclase cumulate should have extra europium (a

---

[*] The concentrations of rare earths and other trace elements in most geological samples are very low, generally measured in parts per million. These can be determined with high precision using an analytical method called *neutron activation*. Weighed samples are placed in a reactor, where they are bombarded with neutrons. These combine with elements in the sample to produce new isotopes, many of which are unstable and have short half-lives. Subsequent decay of these new radionuclides produces gamma (γ) radiation. Gamma rays produce electrons in a detector that are counted when they produce pulses of light (scintillations) in certain crystals. The energy and quantity of the gamma radiation are functions of the identity and amount of each radionuclide, which in turn are used to calculate the concentration of the trace element from which the radionuclide was made.

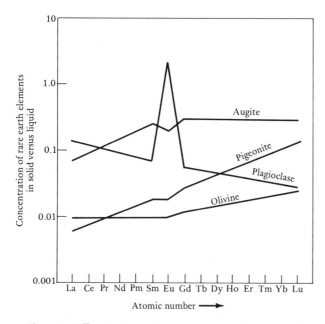

Fig. 4.3. Different minerals incorporate trace elements, such as the rare earths, into their crystal structures in varying amounts as they crystallize from a magma. The experimentally determined rare earth element patterns for four minerals, compared with the abundances of rare earths in the liquid from which they crystallized, are shown in this diagram. The rare earths are plotted according to increasing atomic number. Ratios of less than 1.0 indicate higher values in liquid than in solids, because all of these minerals exclude rare earth elements from their crystal structures to some degree. The one exception is plagioclase, which allows europium (Eu) to replace calcium in its lattice. Understanding these patterns permits the reconstruction of a rock's crystallization history from its trace element abundances in some cases.

positive anomaly), and the melt should have a negative europium anomaly. These elements can also be used to model partial melting. Using the information in Figure 4.3, the mineralogy of the source region and the degree of partial melting can be inferred from the rare earth concentrations of magmas derived from it.

THE EUCRITE ASSOCIATION

The name **eucrite** comes from the Greek "eukritos," meaning "easily distinguished." These meteorites may be easily distinguished from chondrites, but they look superficially enough like terrestrial basalts that they are rarely recovered except from observed falls. Nevertheless, there are some significant differences in the mineralogy of eucrites and terrestrial basalts. The plagioclase in eucrites is rich in calcium and contains very little sodium, and the pyroxene is pigeonite, with composition of approximately $MgFeSi_2O_6$ and very little calcium. Terrestrial basalts generally contain more sodic plagioclase and the calcium-rich pyroxene augite. Another difference is that eucrites contain no water, whereas terrestrial basalts commonly have some hydrous minerals. Eucrites are also reduced, that is, all the iron in them is Fe^{2+} or iron metal, whereas more oxidized terrestrial basalts contain Fe^{2+} and Fe^{3+}. The implication from these differences is that the source region for eucrites was distinct in composition and oxidation state from the earth's mantle, the source region for terrestrial basalts.

Diogenites are composed almost entirely of calcium-poor pyroxene, with only small amounts of plagioclase or olivine. Their mineralogy and oxidation state are similar to those in eucrites, suggesting that the two types of achondrites are related. Moreover, eucrite and diogenite fragments are commonly mixed together to form breccias, supporting the idea that they formed together on the same parent body.

The relationship between eucrites and diogenites can be inferred from an examination of their textures and mineral compositions. The eucrites are commonly fine-grained rocks in which elongated plagioclase grains are enclosed by pyroxenes. This texture also occurs in terrestrial volcanic rocks, indicating that such eucrites were surficial flows. Some eucrites consist of equant, interlocking grains that formed by recrystallization in the solid state (that is, metamorphism). Eucrite layers buried and heated by subsequent volcanic flows might experience this kind of recrystallization. In contrast to these textures, diogenites consist of larger, interlocking crystals, as appropriate to plutonic rocks. The abundance of pyroxene in diogenites is presumably due to fractional crystallization in eucritic magma chambers. Pyroxenes in diogenites are more magnesium-rich than those in eucrites, lending support to the idea that they crystallized early and accumulated. Microscopic views of these contrasting textures, magnified to the same scale, are shown in Figure 4.4.

Most eucrites are actually breccias, composed of angular clasts of volcanic rocks cemented by pulverized mineral grains (see Figure 4.1). Random impacts into the eucrite parent-body surface were bound to excavate and mix in some plutonic diogenite. Achondrites containing both components are called **howardites**. Chemical studies also demonstrate that howardite compositions lie on mixing lines between eucrites and diogenites.

Fig. 4.4. In general, the faster a magma cools and crystallizes, the smaller the size of its crystals will be. Shown here are microscopic views of thin sections of the Pasamonte (New Mexico) eucrite (above) and the Johnston (Colorado) diogenite (below). The fine-grained texture of the eucrite indicates that it is a volcanic rock, and the coarser grain size of the diogenite suggests that it is plutonic.

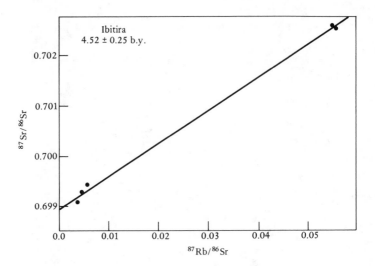

Fig. 4.5. Measurement of the abundances of rubidium and strontium iso-topes provides a means of determining the time of crystallization of igneous rocks. The points in this diagram represent individual minerals separated from the Ibitira (Brazil) eucrite. The slope of the line drawn through these points is related to the age of the meteorite, which in this case is 4.52 billion years. Ig-neous activity clearly occurred very soon after the formation of the eucrite parent body.

Can we tell when igneous activity took place on the eucrite par-ent body? Magmas do not persist in a molten state for very long, and crystallization into solid rocks sets their radioactive clocks ticking. Dating of most eucrites and diogenites using the rubidium-strontium system and other similar isotopic methods gives crystal-lization ages between 4.5 and 4.6 billion years, as illustrated in Figure 4.5. Thus, melting on the eucrite parent body was a very early event in the solar system, roughly contemporaneous with the formation and metamorphism of chondrite parent bodies. Younger radiometric ages measured from a few eucrites can be dismissed as isochrons disturbed by metamorphic recrystallization. Conse-quently, igneous activity on the eucrite parent body not only oc-curred very early but also was short-lived. The rapid heating nec-essary to accomplish this was probably caused by radioactive decay of ^{26}Al. We have already seen that this short-lived isotope may also have caused the metamorphism of chondrite parent bodies, and it may be plausible to view the partial melting event that pro-duced eucrites as metamorphism carried to the extreme.

This possible connection between eucrite melting and chondrite metamorphism would be strengthened if it could be shown that other similarities exist between their parent bodies. Melting of the eucrite source region occurred so early that it seems unlikely that the body was internally differentiated into regions of different compositions before this event. Thus, any reconstruction of the composition of the eucrite source region is probably also a fair estimate of the composition of the whole eucrite parent body. Let us examine how one goes about studying eucrites to see what minerals were left behind in the source region.

To illustrate how this might be done, consider a pan of warm water with white crystals, possibly of salt or sugar, on the bottom. If one were given a sample of water poured from the pan and asked to identify the crystals left behind in the pan, how could this be done? One could, of course, taste the water, but besides being somewhat risky, this might give ambiguous results if the crystals did not impart a distinctive taste to the water or were a mixture of several substances. Another way would be to cool the water, lowering the solubility of the dissolved solid materials and causing more white crystals to precipitate. These crystals might then be identified by other means, such as by their crystal form or chemistry. The crux of this illustration is that the unknown crystals left behind in the pan are the same as those that crystallize from the water on further cooling.

The identity of minerals in a source region can be studied in the laboratory in a somewhat similar manner by remelting a volcanic rock in a furnace and directly observing its crystallization during cooling. Primary liquids that formed by partial melting of several minerals in the source region should still be in **equilibrium** with those minerals at the rock's melting point. If, at the onset of crystallization, a melt can be shown to form a number of minerals simultaneously, these minerals probably constituted a major part of the source region. Such experiments on eucrites indicate that three minerals – olivine, pigeonite, and plagioclase – all generally begin to crystallize within a very narrow temperature interval of less than 10 degrees. The implication is that these three minerals were residues left behind in the eucrite source region and thus must have been important constituents of the original source rocks. This is, of course, true only if eucrites represent primary liquids that have not experienced much fractionation. A further important point is that these experiments were carried out at atmospheric pressure. In similar experiments on terrestrial rocks, the depth of the source region may be estimated by adjusting the pressure higher

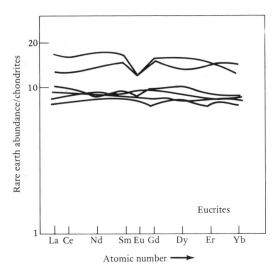

Fig. 4.6. The rare earth abundances in chondrites define the cosmic abundances of these elements. Any deviation from chondritic rare earth element abundances indicates the operation of some fractionation process. It is therefore convenient to divide the analyses of rare earths in a rock by the chondritic abundances so that fractionation can be readily noted. As seen in this figure, eucrites have nearly flat rare earth patterns, so they must have been produced by partial melting of unfractionated source regions similar in composition to chondrites. From such data it has been argued that the eucrite parent body was a chondrite parent body that was heated to the point of partial melting.

until a number of minerals begin to crystallize simultaneously. The fact that eucrites are in equilibrium with three minerals at low pressure implies that the eucrite parent body was small and so would have low internal confining pressures.

Although experiments suggest that olivine, pigeonite, and plagioclase formed the bulk of the eucrite source region, they do not specify the relative proportions of these minerals. The same liquid could be in equilibrium with a residue containing 99 percent olivine crystals or only 1 percent. Luckily, other chemical imprints on magmas can be used to set limits on the proportions of these minerals. The patterns of rare earth elements in several eucrites are shown in Figure 4.6. Plotted in this diagram is the ratio of the abundance of each element to that in chondrites. Chondrites, of course, have cosmic abundances of rare earths, so this ratio readily shows whether any igneous processes have concentrated or depleted these elements. A knowledge of the quantitative behavior of rare earth elements in various minerals can constrain the composition of the eucrite source region. Computations of the rare earth element concentrations that would be produced by partially melting various mixtures of plagioclase, olivine, and pyroxene have been compared with rare earth analyses of eucrites. From this comparison it has been concluded that eucrite magmas were produced by 4 to 15 percent partial melting of **peridotite**, a rock composed mostly of olivine and pyroxene with only a minor amount of plagioclase. This is an interesting result, because peridotite is also thought to compose most of the earth's mantle, the source region for terrestrial basalts. However, the eucrite source region was different in the compositions of the minerals composing the peridotite, as well as oxidation state.

Ordinary chondrites also fit the mineralogical description of peridotites, and they are less oxidized than terrestrial mantle rocks. The peridotite composition of the eucrite source region is likely also to represent the bulk composition of the eucrite parent body, because there was no time for differentiation prior to this early melting event. It appears that the eucrite parent body may originally have been a chondrite parent body that was heated to the point of partial melting.

THE SHERGOTTITE ASSOCIATION

Shergottites take their name from a meteorite that fell in 1865 in Shergotty, located in the Indian state of Bihar. They are basaltic rocks composed primarily of pyroxene and plagioclase, but unlike the eucrites, these meteorites are mineralogically similar to terrestrial basalts. There are two pyroxenes, calcium-rich augite and calcium-poor pigeonite, and the plagioclase consists of approximately equal parts calcium- and sodium-rich end members. These achondrites are also oxidized, containing some ferric iron (Fe^{+3}) in the form of the mineral magnetite. Moreover, like terrestrial basalts, the shergottites contain small amounts of a hydrous mineral, in this case the amphibole kaersutite. In fact, this group of meteorites and their relatives are the only achondrites to contain water bound into crystal structures.

Most shergottites are cumulates, consisting of crystals that probably accumulated at the bottom of a magma chamber. The evidence for this assertion comes from the texture of these achon-

Fig. 4.7. The orientation of pyroxene grains in the Shergotty (India) meteorite suggests that these crystals settled to the floor of a magma chamber. The sketch at the left illustrates how such a preferred orientation of grains is produced. The Shergotty slab at the right has been oriented so that its elongated pyroxene grains are aligned more or less horizontally, as they were when the crystals settled out of the magma more than a billion years ago. Photograph courtesy of the Smithsonian Institution; sketch reprinted from "Basaltic Meteorites" by H. Y. McSween, Jr. and E. Stolper, Scientific American, June 1980.

drites: Elongated pyroxene crystals, shaped more or less like corn cobs, have formed an oriented network, as shown in Figure 4.7. Such elongated grains could hardly be expected to stand on end after sinking to the chamber floor.

Other meteorites apparently from the same parent body are the **nakhlites**, also cumulates consisting mostly of augite, and one **chassignite**, a unique cumulate meteorite from Chassigny (France) containing mostly olivine. These meteorites, too, are oxidized and contain hydrous minerals.

The chemical composition of the shergottites also links them with terrestrial basalts. Figure 4.9 compares the trace-element composition of average shergottite with that of average terrestrial basalt. Elements that fall along the diagonal line show an exact correspondence in these two kinds of rocks. The chemical similarities are striking. No other kind of extraterrestrial sample – eucrite,

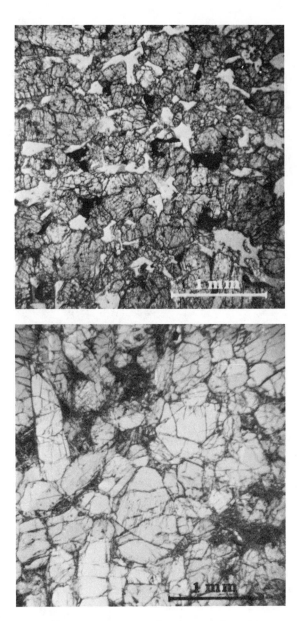

Fig. 4.8. The shergottite association contains a number of meteorites that are very different in appearance, as illustrated by these microscopic views of thin sections of the Zagami (Nigeria) shergottite (above) and the Nakhla (Egypt) nakhlite (below). Both kinds of meteorites are cumulate rocks.

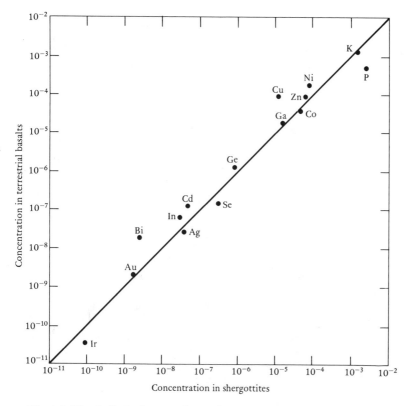

Fig. 4.9. The similarity between shergottites and terrestrial basalts extends to trace elements, suggesting that the source region that partially melted to produce shergottite magmas bears a strong resemblance to the earth's upper mantle, the source region for terrestrial basalts. Elements that plot along the diagonal line have the same abundances in both kinds of rock. This correlation is remarkable because the elements shown here have drastically different geochemical behavior patterns.

chondrite, lunar basalt – has a chemical composition so like those of basalts from the earth. The trace elements plotted in this diagram exhibit a wide range of geochemical behavior, but all are contained in both these materials in approximately the same concentrations. The inference is that the source region that partially melted to produce shergottite magmas must have been very similar in composition to the earth's mantle.

The patterns of rare earth elements in several shergottites are shown in Figure 4.10. The ratio of the abundance of each element to that in chondrites again is plotted in this diagram. Notice that the rare earth pattern for the shergottites is shaped somewhat like

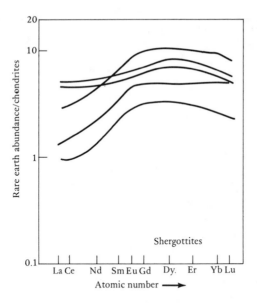

Fig. 4.10. The rare earth element abundances of shergottites, when divided by those in chondrites, are very different from the flat patterns seen in eucrites. These strongly fractionated patterns indicate that the source region for shergottites was not chondritic. Detailed calculations suggest that several periods of melting were required to produce such unusual patterns. The implication of these data is that the shergottite parent body had a complex history of igneous activity.

a lazy S. This is very different from the flat pattern measured in eucrites. Rare earth patterns in partial melts generally mimic the patterns in their source rocks. If the eucrites were formed by partial melting of chondrite, we would expect them to have a flat rare earth pattern similar to that of their source rocks. But how do we explain the shape of the shergottite pattern? It is not an easy task to separate rare earths from one another to form such a skewed pattern; clearly the shergottite source region was not chondritic in composition. Detailed modeling of this pattern indicates that it could be produced only by several generations of melting and crystallization. Therefore, the evolution of the shergottite parent body appears to have been more complex than the single-stage melting event envisioned for the eucrites.

Shergottites have been severely modified by shock metamorphism. Various minerals are affected by shock in different ways, but one of the most interesting responses is the transformation of one mineral into another with a different (usually more compact) crystal structure. Several minerals with the same chemical composition but different crystal structures are called **polymorphs**. Dense polymorphs of olivine and pyroxene have been found in one shergottite. The most obvious mineral transformation that has occurred in all shergottites is the alteration of plagioclase into maskelynite, a glass formed in the solid state. This phase is not really a mineral, because it has no orderly atomic structure. Experiments have produced maskelynite by subjecting plagioclase to shock pressures of at least 30 gigapascals (300,000 atmospheres). Some shergottites also contain veins of shock-melted rock, requiring shock pressures exceeding about 45 gigapascals. The Chassigny meteorite also shows shock effects, but nakhlites do not.

The time of crystallization of the shergottites is their greatest surprise. However, this turns out to be a rather difficult number to obtain, because shock metamorphism has reset most of the radiometric clocks. One isotopic system involving the decay of an isotope of the rare earth element samarium (^{147}Sm) into neodymium (^{143}Nd) may have survived intact. This isotopic clock employs different elements but functions the same way as the rubidium-strontium clock discussed earlier. The shergottites yield a samarium-neodymium isochron corresponding to a time of crystallization approximately 1.3 billion years ago. Several different isotopic clocks in the unshocked nakhlites record a similar time of crystallization. This is a very young age compared with that for eucrites and indicates that igneous activity on the shergottite parent body was much later and may have persisted for a long period of time. Some other isotopic studies suggest even more recent crystallization ages for shergottites; the problem is unresolved, but these are clearly young rocks. The rubidium-strontium isotopic system, having been reset by shock, indicates an age of 180,000 years. This time possibly marks an important impact event in the history of shergottites.

The view of the parent body interior gleaned from shergottites and associated meteorites is rather different from that for eucrites. What we have seen is a geologically complex body, characterized by multiple periods of igneous activity. Isotopic data suggest that the body was originally differentiated about 4.5 billion years ago. The mantle thus formed had a non-chondritic composition, at least in terms of rare earth elements and probably in terms of other elements as well. This source region was remelted about 1.3 billion years ago and possibly later to produce shergottite parent magmas. We cannot reconstruct the exact composition of the source rocks, because the shergottites have experienced fractional crystallization. However, the remarkable correspondence in trace element contents, the similarity in oxidation state, and the presence of water all suggest that the source region for shergottites was very much like that for terrestrial basalts.

METEORITES FROM THE MOON

The scarred face of the moon has been ravaged repeatedly by impacts of meteoroids. Is it possible that lunar materials could have been ejected by such impacts? Calculations suggest that more than a billion grams of impact ejecta should be lost from the moon each year, at least a hundredth of which should be swept up by the earth. This is only a tiny fraction, less than 1 percent, of the meteoritic material arriving from space each year, but it is nonetheless significant. If these calculations are correct, lunar rocks should compose some proportion of our meteorite collections. Such meteorites would certainly be classified as achondrites because of the igneous processes that have dominated the moon's evolution.

The hypothesis that meteorites are lunar samples was proposed more than three centuries ago, and it has been revived many times in the ensuing years. The return of lunar samples by Apollo astronauts and by unmanned Luna landers finally provided an absolute test for this hypothesis. However, from studies of these rocks it quickly became clear that no achondrites had come from the moon, and in the 1970s this idea once again passed into scientific oblivion.

Then, in 1982, fate intervened. Several American scientists were traversing an ice field near the Allan Hills, Antarctica, on snowmobiles. Their purpose was to examine the configuration of the ice sheet; they were not even looking for meteorites. Visibility was poor because of blowing snow, and it was only by the sheerest chance that one of them spotted a small, plum-sized meteorite in his path. The finder, having had considerable experience collecting Antarctic meteorites, recognized that the specimen was different from others he had seen, and he collected it. That evening the weather deteriorated rapidly, making further fieldwork impossible. That chance meteorite was the last specimen collected during the 1981–82 field season in Antarctica.

Some months later a thin section of that meteorite, which had been assigned the name Allan Hills A81005, reached the Smithsonian Institution in Washington. The meteorite proved to be an achondrite breccia. In a preliminary description, the Smithsonian's

Fig. 4.11. The discovery in Antarctica of lunar achondrites is one of the most exciting revelations in meteoritics in the last decade. This photograph shows that the Allan Hills A81005 meteorite is a breccia containing white clasts in a darker matrix. Similar regolith breccias were returned from the moon by the Apollo missions. The cube measuring 1 centimeter on a side is for scale. Photograph courtesy of NASA.

foremost meteoriticist suggested that "some of the clasts resemble the anorthositic clasts described in lunar rocks." This guarded statement was exciting enough to other meteoriticists to generate a flurry of activity, and by year's end no less than 22 research groups were working on this small sample. In early 1983, a special session on ALHA 81005 was convened to air the results at the Fourteenth Lunar and Planetary Science Conference in Houston, Texas. Every participant in the session concurred that the meteorite was probably (most said certainly) of lunar origin. Such unanimity of scientific opinion was startling, even allowing for the fact that it was St. Patrick's Day and spirits were high. Apparently at least one meteorite had come from the moon.

As first noted in the preliminary description, the white clasts in ALHA 81005 do resemble some lunar rocks. These are **anorthosites**, igneous cumulate rocks composed almost wholly of plagioclase, with only minor olivine and/or pyroxene. Also present are dark clasts of basalt similar to those returned from the moon. Many of the clasts are recrystallized or even partly melted, and these are

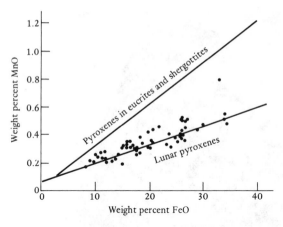

Fig. 4.12. One distinctive fingerprint for lunar rocks is the ratio of manganese oxide (MnO) to iron oxide (FeO) in pyroxenes. These minerals define a trend distinctly different from that of pyroxenes in eucrites and shergottites, as illustrated in this diagram. Analyses of pyroxenes in the Allan Hills A81005 achondrite, shown by dots in the diagram, clearly plot along the lunar trend. This is one piece of evidence that this meteorite came from the moon.

mixed with an assortment of broken mineral grains and glass fragments into a brown glassy matrix. These components, plus clumps of glass-bonded grains (agglutinates), identify this specimen as a regolith breccia. Similar samples were scattered over the lunar surface at the highlands sites visited by Apollo missions.

A more definite indication of lunar origin for this meteorite was provided by the ratios of manganese oxide (MnO) to iron oxide (FeO) in ALHA 81005 minerals. The partitioning of these oxides in pyroxenes is significantly different in lunar rocks and in other achondrites. As illustrated in Figure 4.12, ALHA 81005 pyroxenes fall along the lunar trend. The oxygen isotopic composition of this meteorite is also characteristic. On a diagram of $^{17}O/^{16}O$ versus $^{18}O/^{16}O$, ALHA 81005 plots along the terrestrial fractionation line, as do other lunar samples, but not other achondrites. Finally, this meteorite contains appreciable quantities of noble gases like helium, neon, argon, krypton, and xenon. These gases were implanted by the solar wind into the matrix of the meteorite while it was in a regolith. In contrast to breccias from the moon, even the most gas-rich achondrite breccias have lower concentrations of gases and are relatively depleted in the heavier isotopes.

Subsequent to this revelation that a lunar meteorite had been discovered in Antarctica, Japanese scientists have recognized several other achondrites of lunar origin in their Antarctic collections. These samples, found in the Yamato Mountains, are also regolith breccias. However, each of these and ALHA 81005 are different from one another and must represent different falls.

UREILITES

On the morning of September 10, 1886, several meteorites fell near the village of Novo Urei in the Krasnoslobodsk district of Russia. This was a particularly interesting fall for several reasons. One of these stones was soon recovered by local peasants, whereupon it was broken apart and eaten. The motivation for this rather unusual action is not known, but this constituted an impressive feat from a dental perspective, because the meteorite contained numerous small diamonds. The uneaten specimens from this fall proved to be a unique type of achondrite; subsequently recovered meteorites of this type are known as **ureilites**.

The ureilites are arguably the most bizarre and perplexing of all achondrites. They consist principally of the minerals olivine and pigeonite. Filling spaces between the larger silicate grains is a matrix of graphite or diamond, iron-nickel metal, troilite, and other minor phases. A microscopic view of a ureilite is shown in Figure 4.13. Where carbonaceous matrix is in contact with olivine or pyroxene, the two have reacted. Graphite caused Fe^{+2} in the rims of olivine and pigeonite crystals to be reduced to iron metal, forming correspondingly magnesium-enriched compositions at the edges of grains.

The coarse grain size of ureilites indicates that they are plutonic rocks, but it is disputed whether they formed from accumulations of minerals in a magma chamber or are residual crystals left over from partial melting. Olivine and pyroxene grains in ureilites exhibit preferred orientations, an observation most consistent with the cumulate hypothesis, but the point is still contested.

Another unresolved question is the source of the carbon in ureilites. Proponents of the cumulate origin favor a purely igneous model in which the carbonaceous matrix represents crystallized melt trapped between accumulating crystals. Advocates of the residue origin suggest that the carbonaceous matrix is unrelated to the olivine and pyroxene network, having been forcefully injected into the rock at some later stage.

One of the most interesting characteristics of ureilites (besides their possible tastiness) is that they have experienced varible but

Fig. 4.13. The ureilites are the least understood of the achondrites. This microscopic view of a thin section of the Kenna (New Mexico) ureilite shows olivine crystals with preferred orientations in a background of opaque graphite and other phases. Photomicrograph courtesy of J. L. Berkley (State University of New York, Fredonia).

typically intense shock metamorphism. In many specimens, graphite, the original carbon mineral, has been partly transformed by shock into its polymorphs diamond and lonsdalite. The impact process, which caused shock metamorphism, has also been argued to have caused injection of the carbonaceous matrix. The chemistry of ureilites is very strange and might require some kind of mixing process. For example, the rare earth pattern for ureilites is V-shaped. Enrichment of both heavy and light rare earth elements is very difficult to achieve by any igneous process, and mixing of light rare earth-enriched and heavy rare-earth-enriched materials by shock could explain this pattern. Shock mixing probably should produce veins of carbonaceous matrix that crosscut olivine and pigeonite; however, the matrix is confined to the areas between larger silicate grains and occurs even in relatively unshocked ureilites.

The presence of carbon minerals may suggest some ultimate link between ureilites and carbonaceous chondrites. The oxygen isotopic compositions of these two meteorite groups are similar, and the relative abundances of noble gas isotopes in ureilites are close to values measured in some carbonaceous chondrites. However, the nature of the connection, if any, between these meteorite classes is unknown. The ureilites remain a big mystery.

ACHONDRITE AFFILIATIONS

We have now described the eucrite and shergottite associations, the lunar achondrites, and the ureilites. This listing does not exhaust the known achondrite types, but it is sufficient to illustrate some differences among igneous meteorites. We have also seen that achondrites formed by igneous processes on the same parent body can be different. How many parent bodies are necessary to have produced these achondrites?

One way to recognize possible parent-body affiliations between different achondrites is to study which types occur together in breccias. Howardites provide a good example of using this approach to relate eucrites and diogenites. An entirely different way is by comparing their oxygen isotopic compositions. In a previous chapter we mentioned how igneous processes such as partial melting and fractional crystallization can smear isotopic compositions along a line of slope $+\frac{1}{2}$ on a plot of $^{18}O/^{16}O$ versus $^{17}O/^{16}O$. The isotopic compositions of igneous rocks are limited by the fact that they can lie only along a mass fractionation line that passes through the initial parent-body composition. Oxygen isotopic compositions for the achondrites that have just been discussed are shown in Figure 4.14. The lunar achondrite, ALHA 81005, falls along the terrestrial mass fractionation line, as do Apollo and Luna samples. The earth-moon data demonstrate that two (or more?) bodies can have the same mass fractionation line. However, it seems probable that achondrites, like chondrites, are samples from only a few parent bodies. The grouping of achondrites into associations related by igneous processes also lends credence to this idea. Eucrites, diogenites, and howardites lie along one mass-fractionation line, and shergottites, nakhlites, and the only known chassignite define another. This does not prove that all of the members of these two associations formed on only two parent bodies, but it seems likely. The few ureilites that have been isotopically analyzed plot near the carbonaceous chondrite mixing line. This may reflect admixture of carbonaceous material that was isotopically similar to carbonaceous chondrites.

We must conclude that these four meteorite associations probably sampled four parent bodies. The variability of achondrite types on one parent body can be understood as a natural and expected

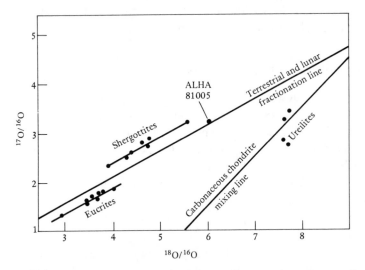

Fig. 4.14. Various classes of achondrites define distinct mass fractionation lines in terms of their oxygen isotopic compositions. Because meteorites cannot move from one fractionation line to another by igneous processes such as melting or crystallization, each line presumably represents a different parent body. Allan Hills A81005 plots on the terrestrial and lunar fractionation line, another piece of evidence supporting its lunar origin. Ureilites curiously fall along the carbonaceous chondrite mixing line, possibly suggesting some mixture of chondritic and achondritic materials.

consequence of igneous processes. There are other kinds of achondrites we have not considered, most notably the aubrites, which probably formed by melting on an enstatite chondrite-like parent body. The list also includes some unique meteorites that by themselves do not provide enough information to say much about their parent bodies. All of these taken together would possibly only double the number of required parent bodies for achondrites.

A PERSONAL TOUCH

In the National Air and Space Museum in Washington, DC, a metal pylon stands just below the Wright brothers' airplane. Embedded in the pylon is a small black rock shaped like an arrowhead. People, young and old, approach it under the eye of a guard. One by one they advance, reach out their hands, touch the rock, and then walk slowly away, smiling or thoughtful. They have just done something that was once impossible. They have touched a piece of the moon. (From "Getting Our Hands on the Universe" by B. M. French, *The Planetary Report*, March 1984.)

What was so nicely described here is an intensely human response, understandable even by scientists who handle lunar samples routinely. The touching of something so foreign as a rock from the moon is an experience not easily bypassed.

Achondrites elicit the same kind of wonder (at least in those few who know what they are), because they also sample geologically processed worlds. In this chapter we have described four associations of achondrites. The meteorites within each association may be related through fractional crystallization, degree of partial melting, or non-magmatic processes such as impact mixing.

Despite the fact that most achondrites are very different in composition from terrestrial igneous rocks, they appear to have formed by parallel processes. For this reason, achondrites seem comfortingly familiar to many geologists, at least in comparison with chondrites for which we have no terrestrial analogues. Correctly interpreting the evidence in achondrite associations is a difficult task, plagued by assumptions and fraught with ambiguities. However, the outcome – glimpses into the geologic evolution of other worlds – is certainly worth the effort. In the following chapter we shall attempt to define more fully the characteristics of each achondritic parent body and to identify them where possible.

SUGGESTED READINGS

There are very few nontechnical papers on achondritic meteorites. Many of the works cited here require some knowledge of the specialized vocabulary of geology.

GENERAL

Dodd R. T. (1981) *Meteorites: A Petrologic-Chemical Synthesis*, Cambridge University Press, Cambridge, England. 368 pp. (Chapters 8 and 9 provide excellent technical discussions of achondrites.)

Basaltic Volcanism Study Project (1981) *Basaltic Volcanism on the Terrestrial Planets*, Pergamon Press, Oxford, 1,286 pp. (The definitive technical work on volcanism; contains a review chapter on achondrites.)

PETROLOGY OF EUCRITES

Bunch T. E. (1975) Petrography and petrology of basaltic achondrite polymict breccias (howardites). In *Proceedings of the 6th Lunar Science Conference*, Pergamon Press, New York, pp. 469–492. (Technical paper describing brecciated members of the eucrite association.)

Stolper E. (1977) Experimental petrology of eucritic meteorites. *Geochimica et Cosmochimica Acta* 41, 587–611. (Technical paper that utilizes crystallization experiments to understand the igneous processes that produced eucrites.)

PETROLOGY OF SHERGOTTITES

Stolper E. and McSween H. Y. Jr. (1979) Petrology and origin of the shergottite meteorites. *Geochimica et Cosmochimica Acta* 43, 1475–1498. (Technical paper that describes the characteristics of shergottites and suggests processes by which they formed.)

AGES

Birck J. L. and Allegre C. J. (1978) Chronology and chemical history of the parent body of basaltic achondrites studied by the ^{87}Rb-^{87}Sr method. *Earth and Planetary Science Letters* 39, 37–51. (Technical paper that presents radiometric ages for eucrites.)

Shih C. Y., Nyquist L. E., Bogard D. D., McKay G. A., Wooden J. L., Bansal B. M., and Wiesmann H. (1982) Chronology and petrogenesis of young achondrites, Shergotty, Zagami, and ALHA 77005: Late magmatism on a geologically active planet. *Geochimica et Cosmochimica Acta* 46, 2323–2344. (Technical paper that describes radiometric age data and interpretations for shergottites.)

PETROLOGY OF UREILITES

Berkeley J. L., Taylor G. J., Keil K., Harlow G. E., and Prinz M. (1980) The nature and origin of ureilites. *Geochimica et Cosmochimica Acta* 44, 1579–1597. (Technical paper that presents data on ureilites.)

LUNAR ACHONDRITES

Marvin U. B. (1984) A meteorite from the moon. *Smithsonian Contributions to the Earth Sciences* 26, 95–103. (Nontechnical paper that summarizes the properties of achondrite ALHA 81005 and evidence for a lunar origin.)

RELATIONSHIPS AMONG ACHONDRITES

Clayton R. N. and Mayeda T. K. (1983) Oxygen isotopes in eucrites, shergottites, nakhlites, and chassignites. *Earth and Planetary Science Letters* 62, 1–6. (Technical paper that presents oxygen isotopic data for various achondrite groups.)

5 Achondrite parent bodies

Apollo – the name still evokes a special feeling in anyone old enough to remember sitting riveted to the television to watch this "one giant leap for mankind." The series of missions that culminated in placing men on the moon ranks as one of the triumphs of human endeavor. The early Apollo flights allocated little time to sample collection because of concern about man's ability to cope with the harsh lunar environment, but on later missions extended sampling forays, complete with transport vehicles, were carried out with more confidence. A total of 381 kilograms of lunar rocks and soil were ultimately returned to earth by the Apollo astronauts. Assuming a cost of $24 billion for the total Apollo program, this works out to $28,500 per pound of sample. From a scientific perspective, these are bargain-basement prices. Even from a purely economic point of view, the most cynical (but informed) critics must acknowledge that the technological spinoffs have more than compensated for the cost of this program.

Consider how much more of a bargain are achondrites. Celestial traffic accidents have provided geologically processed samples of other worlds at no cost and with little effort on our part. But we pay another kind of price in trying to use these meteorites for scientific research: For most achondrites we do not know with absolute certainty from which parent bodies they are derived. Nevertheless, as we shall discover in this chapter, we can make some very informed guesses about the identities of many achondrite parent bodies.

OUR NEAREST NEIGHBOR

Because we have already mentioned the Apollo program, let us start with the best known achondrite parent body, the moon. The lunar achondrites are clearly exceptions to our generalization about the uncertainty in identifying achondrite parent bodies, but only because we have Apollo samples, as well as those collected on the

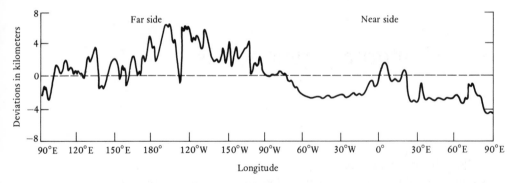

Fig. 5.1. The moon's surface can be divided into areas of ancient crust with rough topography and relatively smooth, low-lying areas covered by flood basalts. This topographic profile was made by a laser altimeter on the Apollo 16 spacecraft as it orbited the moon. Data are shown as deviations from a sphere of average lunar radius, corresponding to 1,738 kilometers. The far side consists wholly of highlands. The low areas on the near side are maria.

unmanned Luna missions of the Soviet Union, with which to compare them. What has been learned about the moon? The following picture has been pieced together from petrologic, geochemical, and isotopic investigations of lunar samples, as well as high-resolution photographs and remote-sensing measurements of the lunar surface made from orbiting spacecraft.

Even casual inspection of photographs of the near side of the moon reveals that the lunar landscape can be divided into two kinds of areas. The **highlands** are higher, rougher terrains of lighter color than the **maria**, which are dark, relatively smooth lowlands. The far side of the moon, which always faces away from the earth and was never seen until the advent of spacecraft, consists only of highlands terrain. The contrast in topographic elevation between highlands and mare regions is shown in Figure 5.1, a laser altimeter traverse made as the Apollo 16 orbiter circled the moon. All of the Apollo and Luna missions landed on the near side, and most sampled the maria, for the obvious reason that smoother surfaces provided safer landing sites. These sampling sites are pointed out in the near-side photograph in Figure 5.2.

The rugged nature of the highlands is attributable to ancient meteoroid impacts that formed countless overlapping craters. Some of these craters are huge, spanning hundreds of kilometers from rim to rim. Such intense cratering spread highlands ejecta over much of the lunar surface. Particles of plagioclase-rich rock found

Fig. 5.2. This photograph of the near side of the moon clearly shows the distinction between the rough, heavily cratered highlands that are light in color and the relatively smooth, dark maria. The arrows mark the landing sites of Apollo and Luna missions that returned lunar samples to earth. Courtesy of the Lunar and Planetary Institute.

in soil collected during the Apollo 11 mission (a mare landing site, and the first sample return mission) led some investigators to speculate that the highlands consisted of anorthosite. That such rocks are major highlands components was subsequently confirmed by the Apollo 15 and 16 missions that actually landed in the highlands. Many highlands rocks also contain appreciable amounts of

Fig. 5.3. Much of the lunar highlands consists of coarse-grained anorthositic rocks that formed by flotation of plagioclase feldspar in a deep magma ocean. This specimen, composed of white plagioclase grains and dark olivine, formed 4.5 billion years ago. It is rare to find such well preserved pieces of the ancient lunar crust in the rock collections brought back from the moon, because meteorite bombardment has broken most of them into tiny fragments. The cube is 1 centimeter on a side. Photograph courtesy of NASA.

olivine and/or pyroxene, although the mineralogy of most of these is dominated by calcic plagioclase. Highlands rocks are completely anhydrous and were formed under highly reducing conditions. Radiometric ages for such anorthositic rocks range from 3.9 to about 4.6 billion years. Most of the isotopic clocks in these rocks have been reset by shock metamorphism, and only the oldest ages indicate the true time of formation of the highlands. Anorthositic rocks form a crustal layer that is on the order of 60 kilometers thick on the near side and as much as 100 kilometers thick on the far side. This crust formed when the moon was extensively melted to form a "magma ocean" in its earliest history. The global magma body, apparently of basaltic composition, experienced fractional crystallization as less dense plagioclase crystals floated to the top and solidified as the feldspar-rich highlands. The difference in thickness of the anorthositic crust on the near and far sides may be related in some way to the gravitational pull of the earth acting continuously on one face. This stage of lunar history had ended by about 4.4 billion years ago.

Soon after solidification, the highlands were invaded by new

Fig. 5.4. The lunar maria are vast lakes of crystallized basalt. This sample is riddled with holes, called vesicles, that formed as dissolved gases in the magma formed bubbles during its ascent. The cube is 1 centimeter on a side. Photograph courtesy of NASA.

pulses of magma. Many of these crystallized to form plutons that contained less plagioclase and more pyroxene than the older anorthositic crust. An unusual type of volcanic rock, called KREEP because of its high contents of potassium (K), rare earth elements (REE), and phosphorus (P), erupted in some areas. By about 4.0 billion years ago, this second period of igneous activity was complete.

Coincident with the formation of the lunar crust was a stage of cataclysmic meteoroid bombardment. The outer several kilometers of the moon were virtually demolished and reduced to a cratered pile of rubble. Heavy battering by large meteoroids stopped rather suddenly at about 3.9 billion years ago, but left in its wake enormous circular basins that were the loci for the next (and final) stage of lunar evolution.

From approximately 3.9 to 3.0 billion years ago, the gigantic basins were filled with vast lakes of basaltic magma that crystallized to form the maria. The **mare basalts** consist mostly of pyroxene, plagioclase, and sometimes olivine. Like the highlands rocks, mare basalts contain no bound water and were formed under highly reducing conditions. Mare basalt magmas probably formed by partial melting of the olivine- and pyroxene-rich cumulate that settled

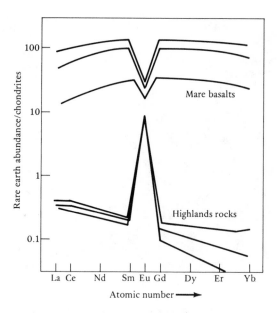

Fig. 5.5. *Trace elements provide important clues to the origin of lunar rocks. The rare earth element patterns for highlands rocks have positive europium (Eu) anomalies (spikes in this diagram), reflecting the accumulation of extra plagioclase from the magma ocean. The complementary olivine- and pyroxene- rich cumulate that must lie below the plagioclase-rich crust therefore has a neg- ative europium anomaly. Mare basalts probably represent partial melts of this underlying cumulate material. As seen in this diagram, they have inherited the negative europium anomaly of their source region.*

from the magma ocean as a complement to the anorthosite crust. The evidence for this assertion comes from measured rare earth patterns of these rocks. The highlands rocks have positive euro- pium anomalies, as illustrated in Figure 5.5, because they consist mostly of cumulate plagioclase that scavenged this element. The plagioclase-depleted residue below must therefore have a negative europium anomaly. Melts derived from this underlying region, the mare basalts, inherited the rare earth pattern of their parent rocks, as shown in the same figure. Mare basalts vary in their contents of the titanium-bearing mineral ilmenite ($FeTiO_3$), and titanium con- tents are commonly used to classify these rocks. Experiments in- dicate that simultaneous crystallization of most minerals occurs at lower pressures for high-titanium basalts than for low-titanium basalts. From such data it has been concluded that low-titanium mare basalts formed by melting at deeper levels than their high- titanium relatives. The ages of low-titanium basalts are also older. The inference is that the zone of partial melting inside the moon

Fig. 5.6. Meteoroid impacts have produced a thick regolith on the surface of the moon. This regolith breccia, returned by Apollo 17 astronauts, is composed of angular clasts and rock powder welded into a coherent rock. This sample is similar in many respects to the breccias formed on asteroid surfaces. Photograph courtesy of NASA.

retreated to deeper levels with time. By about 3.0 billion years ago, the zone of melting had been drawn down so far and the rigid outer portion of the moon had become so thick that magmas generated internally could no longer escape to the surface, and volcanism ceased. The absence of mare basalts on the lunar far side probably reflects the greater thickness of anorthositic crust that had to be traversed.

Although fierce bombardment by giant meteoroids ceased about 3.9 billion years ago, impacts by smaller objects continued throughout the rest of lunar history at a reduced rate. The effect of this continued battering has been to produce a thick regolith composed of unconsolidated rock dust and compacted breccias like that shown in Figure 5.6. This is the same kind of breccia that composes all of the known lunar achondrites.

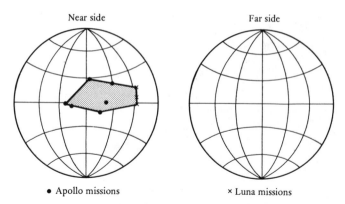

Near side   Far side

• Apollo missions   × Luna missions

Fig. 5.7. Sampling of the lunar surface has been very limited, primarily because relatively smooth, safe landing sites were selected for sample-return missions. A polygon surrounding all of the Apollo and Luna collection sites covers only 4.7 percent of the moon's surface. In all likelihood, lunar meteorites come from outside this small explored area.

With so much lunar material collected by the Apollo and Luna missions, of what scientific value are a few more small regolith breccias? After all, there are reasons to believe that what we have already learned from these missions can be extrapolated to the rest of the moon. Remote-sensing measurements made by several Apollo orbiters (the unlucky astronaut left aloft on each mission actually collected very important data) detected high concentrations of aluminum wherever the spacecraft passed over highlands areas, supporting the idea that they consist of plagioclase-rich (hence aluminum-rich) anorthositic rocks. Nevertheless, there is no substitute for ground truth, and it would be extremely informative to have samples from other parts of the moon.

The surface area bounded by all of the Apollo and Luna landing sites is only a small fraction, 4.7 percent, of the total lunar surface, as illustrated in Figure 5.7. Odds are high that the lunar achondrites sampled regions outside this limited area. Unfortunately we do not know exactly where on the lunar surface they originated, but we may be able to narrow the possibilities. The absence of particle tracks in mineral grains of ALHA 81005 is consistent with a short flight time in space (discussed in more detail in a later chapter). This implies that the impact crater produced when this meteorite was ejected from the lunar surface should be young and relatively fresh. The spectral reflectivity of this achondrite has also been compared with those obtained telescopically from restricted areas of the moon. The meteorite spectrum is unlike that of any area already visited, but it does resemble those of some fairly young craters such as Tycho and Aristarchus. There also remains the possibility that one or more of these meteorites could be from the lunar far side.

The last Apollo mission, Apollo 17, returned with its booty of lunar samples in late 1972. That Antarctic snowmobile excursion during which the ALHA 81005 meteorite was accidentally discovered has been facetiously called "the Apollo 18 mission," because it provided additional lunar samples. The recognition of several other lunar achondrites from Antarctica kindles hope that sampling of the moon will become even more complete.

A MELTED ASTEROID

Chondrites represent primitive solar system materials that have been spared the effects of geologic processing. They apparently avoided this fate because they resided in asteroidal parent bodies too small to have sustained the high temperatures necessary for igneous activity. However, it would be erroneous to conclude from this that no asteroids were ever melted.

We previously examined evidence that the eucrite source region was chondritic in its chemical and mineralogical composition. Because the eucrites are so very old, it is unlikely that their parent body had been differentiated into compositionally distinct layers before the source region partially melted, implying that the whole parent body was chondritic. This suggests an obvious link with the asteroids from which chondrites were derived. The experiments indicating that eucrite magmas were partial melts formed at low pressure are also consistent with an asteroidal source, because small bodies have low internal pressures.

What would cause a particular asteroid to melt while others did not? Let us make the assumption that the heat source for either partial melting on the eucrite parent body or metamorphism on other chondrite parent bodies was the same, which is tantamount to saying that igneous activity is simply an extension of metamorphic processes. It has already been suggested that this heat source was the rapid decay of ^{26}Al, a short-lived nuclide present in the early solar system, but now extinct. If matter were distributed heterogeneously in the solar nebula, asteroids that chanced to incorporate more aluminum (or more ^{26}Al in the case of isotopic heterogeneity) might melt. Even if matter were distributed homogeneously, asteroids that accreted early would contain a greater complement of "live" ^{26}Al and would be capable of experiencing higher temperatures than their cousins that accreted after much of the ^{26}Al had already decayed. Another possibility is that a larger asteroid could sustain higher internal temperatures than a smaller body, because of the inefficiency of heat conduction through rock. Thus, from its very origin the eucrite parent body may have inherited a favorable disposition toward igneous activity, predicated on its composition, time of accretion, or size.

The eucrite parent body was chondritic in composition but in some way probably was not a garden-variety asteroid. It is possible to infer something about its internal structure from the meteorites themselves. The great bulk of this body (90 percent or so) must have been composed mostly of olivine and pyroxenes, the basic mineral constituents of chondrites. Most of this peridotite rock was probably highly metamorphosed and in places had partial melts extracted. Eucrite magmas must have worked their way through this recrystallized interior toward the surface. The ascent of some magma pulses was halted en route as they underwent fractional crystallization to form diogenite cumulates. Other eruptions surged onto the surface or were sandwiched between already solidified flows. The gross stratigraphy of the eucrite parent body is thus proposed to be recrystallized peridotite (the residue from partial melting), succeeded outward by diogenite and finally eucrite, as illustrated in Figure 5.8. Impacts subsequently must have blurred this pattern somewhat by mixing surficial eucrite with excavated diogenite to form howardites, at least in the vicinity of craters.

The dimensions of the interior stratigraphic layers are probably impossible to deduce from meteorite studies, but the thickness of the outside eucrite layer can be estimated from the rate at which the most deeply buried eucrites cooled. The slowest cooling rate naturally corresponds to the most deeply buried rock.

Cooling rates for eucrites have been determined by two different means. In one study, small samples of the Stannern (Czechoslovakia) eucrite were melted and allowed to crystallize at specific cooling rates. Stannern, like many eucrites, is a breccia composed of clasts with varying textures that presumably formed at different depths. Observed changes in the products of the experimental runs, such as the widths of plagioclase crystals, were calibrated with the experimental cooling rates to produce a cooling speedometer applicable to eucrites. It was found that cooling rates ranging from 100 to 0.1°C per hour could produce the range of textures in in-

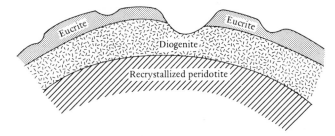

Fig. 5.8. The stratigraphy of the outer part of the eucrite parent body probably looks something like this sketch. This schematic cross section is inferred from the nature of the meteorites themselves. Eucrites are basaltic flows, and diogenites are plutonic rocks that formed at deeper levels. If the parent body has a chondritic bulk composition, as has been argued in the previous chapter, the interior must contain large amounts of peridotite that have been recrystallized and partially melted to produce the parent magmas of eucrites and diogenites. The absence of meteorites consisting of recrystallized peridotite suggests that the eucrite parent body has not suffered catastrophic fragmentation and is still intact.

dividual clasts of the Stannern breccia. Such rapid cooling must have occurred on or very near the surface. Cooling rates have also been deduced from features developed within pyroxene crystals. Pyroxenes are complex solid solutions of calcium-rich (augite) and calcium-poor (pigeonite) end members at high temperatures, but the homogeneous mixture breaks down as temperature decreases. This unmixing process is called **exsolution**. Like an oil-and-vinegar salad dressing that separates into its two constituents when given sufficient time, the components of pyroxenes unmix under slow cooling. The pyroxenes in eucrites consist mostly of the pigeonite component, so calcium ions diffuse through the solid crystals to certain locations at which augite grows. Augite forms flat plates whose orientation is fixed by the crystal structure of the host pigeonite, and whose widths are controlled by the cooling rate. Using the known diffusion rate of calcium in pyroxene, it is possible to calculate the widths of augite exsolution plates grown at different cooling rates. Cooling rates for most eucrites were rapid, but somewhat slower rates were obtained for a few coarser-textured meteorites. The slowest cooling rates determined by this method were about 1°C per 10,000 years, representing the most deeply buried eucrite samples. This study suggests that the thickness of the eucrite layer was no more than about 15 kilometers.

Impacts into the outer stratigraphic layers of the eucrite parent

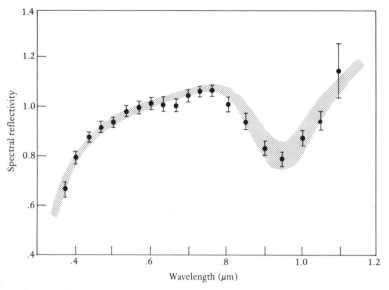

Fig. 5.9. The spectrum of sunlight reflected from the surface of asteroid 4 Vesta (dots with analytical error bars) is very similar to the spectra for eucrites measured in the laboratory (band). The dip in each curve at about 0.95 micrometer corresponds to a strong absorption band of pyroxene. This distinctive spectral match suggests that Vesta is the eucrite parent body.

body have left their imprint in the form of mixing of rocks from different levels. Howardites contain clasts of both eucrites and diogenites, and uncommon olivine grains in these meteorites could even be from the underlying peridotite. A few fragments of carbonaceous chondrite debris have also been noted in howardites, undoubtedly having been derived from impacting meteoroids. The development of a regolith on the eucrite parent body is confirmed by the implantation of solar wind gases and solar flare particles in howardite mineral grains.

LOOKING FOR A NEEDLE IN A HAYSTACK

The hypothetical eucrite parent body we have reconstructed is made predominantly of recrystallized peridotite, not eucrite or diogenite. If the parent body had been destroyed by a large impact, these peridotite meteorites should be relatively common, at least in relation to the number of eucrites and diogenites. However, meteorites matching this peridotite are unknown. It is therefore logical to conclude that the eucrite parent body is still intact. Is it possible

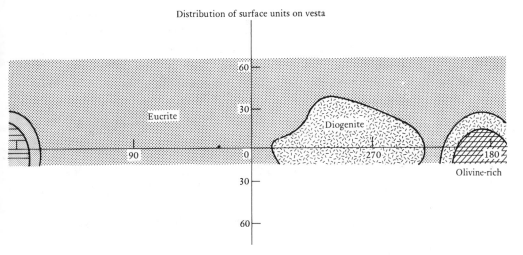

Fig. 5.10. This detailed map of the equatorial region of the surface of 4 Vesta was made by splicing together individual spectra taken as the asteroid rotated. The large circular regions are apparently underlying rocks exposed in the floors of large craters. Vesta's interior is made of rocks similar in mineralogy to diogenites and olivine-rich diogenites (peridotites). This map seems to confirm the stratigraphy of the eucrite parent body inferred in Figure 5.8.

to pinpoint this asteroid among its chondritic neighbors in the asteroid belt? This would seem like a formidable task, but apparently it has been done.

The mineralogy of eucrites provides a rather distinctive reflectance spectrum with a strong absorption feature near 0.9 micrometer attributable to pyroxene. In Figure 5.9, the spectra of several eucrites are compared with that of one particular asteroid, **4 Vesta**. The match is obvious. In fact, the recognition that the spectrum of Vesta was similar to those of eucrites was the first correlation made between any asteroid and meteorites using this observational technique.

More recently, the spectral reflectance technique has been refined in an attempt to construct a map of the surface of Vesta. This is a difficult undertaking, because it is not possible to resolve only a part of the surface of Vesta. However, the spectrum measured at any one time is integrated over only one hemisphere, the side facing the earth. By taking a spectrum at various time intervals, it is possible to see reflectivity differences in different parts of the asteroid as it rotates. When these spectral snapshots are spliced together, a mosaic map of the equatorial region of Vesta is produced. This amazing map is presented in Figure 5.10. What we can see is a eucrite surface punctuated by large, roughly circular areas of different reflectivities. These certainly must be impact craters exposing underlying diogenite on the crater floors. The "olivine-rich diogenite," which apparently lies below the diogenite layer, is not a known meteorite type, but is suggestive of the recrystallized peridotite mantle hypothesized earlier. This work seems to provide elegant confirmation of the parent-body stratigraphy inferred from properties of these meteorites.

Because we know that a regolith formed on the eucrite parent body surface, it may seem surprising that the eucrite and diogenite areas are so cleanly delineated on this map. Howardites document the fact that eucrite and diogenite were mixed in near-surface breccias. The explanation probably lies in the recognition that the spectrum of howardites is intermediate between those of its components, and howardite crater rims would not be recognized at the resolution of this technique.

4 Vesta is the third largest asteroid. Its size has been determined by a variety of techniques, and all give consistent results. For example, its diameter determined from radiometry is 538 kilometers, from polarimetry 558 kilometers, from sparkle interferometry 550 kilometers, and from occultation of a star 549 kilometers. The relatively large size of Vesta probably accounts in some measure for its place of honor as the first asteroid to be linked to a specific kind of meteorite.

One rather surprising result of the spectrophotometric survey of the asteroid belt is that no other asteroids with eucrite-like surfaces have been discovered. The only obvious difference between 4 Vesta and other asteroids is its relatively large size. This plus the absence of eucrite parent bodies among smaller asteroids suggest that size may have been an important if not controlling factor in causing melting on this asteroid. This would be a very convincing argument, except that 1 Ceres and 2 Pallas, which are even larger than Vesta, appear to have chondritic surfaces. The reason that this particular asteroid experienced igneous activity is still unknown.

It is perhaps misleading to suggest that the search for the eucrite parent body among the thousands of other asteroids was like looking for a needle in a haystack. The distinctive spectrum and large size of 4 Vesta made it stand out from the rest. Despite the fact that Vesta may be unique, it remains an amazing achievement that the precise asteroidal parent body for this class of meteorites can be determined with some confidence.

ON A GRANDER SCALE

The success of spectrophotometry in identifying the eucrite parent body leads to the expectation that finding the shergottite parent body should be an equally easy task. The reflectance spectra for vitually every asteroid as bright as Vesta down to 25 kilometers in diameter have now been measured, but no respectable match has been found for shergottites. Could it be that the shergottites were not derived from an asteroid? Perhaps we have been searching on the wrong scale.

One of the loosely coded but obvious patterns of biology is that big things (say elephants and redwood trees) live longer than small things. If a planet or planetesimal can be construed as having a lifetime, it would be based on the length of time the body can sustain internal heat, reflected in the duration of its igneous activity. It is a curious coincidence that, as in the biological analogy, these geologic lifetimes also show a rough correlation with body sizes, as illustrated in Figure 5.11. Igneous processes on asteroid 4 Vesta ceased about 4.5 billion years ago, halting almost as soon as they had begun. Lunar volcanism persisted until approximately 3.0 billion years ago, and volcanic plains on Mercury may be as young as 2.5 billion years. Martian igneous activity lasted at least until 1.1 billion years ago and likely to more recent times. The earth and possibly Venus are still volcanically active. The one known exception to this relationship between planetary size and duration of volcanism is Jupiter's satellite Io, a currently volcanically active body only slightly larger than the earth's moon. The reason for this discrepancy will be discussed later.

It may seem unclear how the ages of planetary volcanic features, from which we have no samples to date radiometrically, have been determined. These are measured by counting the number of craters per unit area in photographs of uneroded volcanic surfaces; the older a surface is, the more impacts it will have accrued. If the cratering rate is known or can be estimated, an absolute age can be estimated from the crater density. The crater production rate has been calibrated for the moon, because we have radiometrically dated samples from the same volcanic terrains for which crater densities have been counted. The meteoroid flux apparently was far from being linear with time, because of the heavy bombardment experienced before about 4.0 billion years ago. Therefore, ages of surfaces must be estimated from graphs that show how

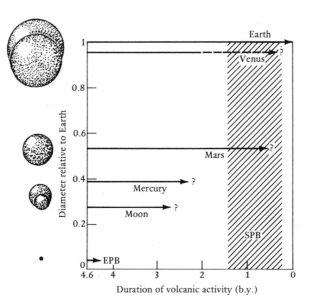

Fig. 5.11. The duration of volcanic activity is apparently related to planet size. In this figure, planetary diameters are shown as fractions of the earth's diameter. The length of time from the beginning of the solar system 4.6 billion years ago until volcanism ceased is illustrated by arrows. Question marks for most planets reflect the uncertainty in ages of volcanic plains determined from measured crater densities. EPB is the eucrite parent body, presumably asteroid 4 Vesta, which supported igneous activity for only a short time after its formation. The band labeled SPB represents the rather uncertain range of radiometric ages for achondrites derived from the shergottite parent body. This late igneous activity suggests that the shergottites may have formed on a large planet like Mars.

crater production rates have changed with time, as illustrated for the moon in Figure 5.12. The cratering rates for other planets are not known precisely, but can be estimated from probabilities of collisions with other bodies versus the moon. Only the grossest Venusian volcanic features can be observed by radar, and their crater densities cannot be counted directly because of the obscuring cloud cover. However, some researchers have argued that Venus is probably volcanically active now, because emissions from active eruptions are suggested by fluctuating contents of sulfurous gases measured in the atmosphere coupled with observed lightning flashes.

The cause of the relationship between planetary scale and duration of volcanism is easy enough to fathom. Volcanism is driven by internal heat, in turn generated by radioactive decay of both

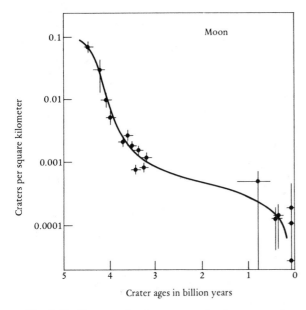

Fig. 5.12. The curve for the crater production rate on the moon was determined empirically by comparing counted crater densities on volcanic plains for which radiometric age determinations of lunar samples were available. These data points are shown by dots with error bars. Using this curve, the age of any lunar terrain can now be determined by counting its crater density.

short- and long-lived radionuclides. Rocks act as thermal insulation that holds in the heat, and the thicker the rock blanket, the longer heat can be retained internally. Small bodies may experience heating for a short time due to ^{26}Al decay, which generates heat faster than it can be conducted away, but they cannot retain the more gradually produced heat from fission of the long-lived radionuclides of uranium, thorium, and potassium. Larger planets, being better insulated, can utilize these lumbering decay schemes to fuel their internal heat engines for billions of years. The explanation of the current volcanic activity on the small Jovian moon Io is that its heat is not internally derived from radioactive decay, but rather is due to a peculiar and continuing gravitational interaction with massive Jupiter.

The reason for this discussion of geologic lifetimes is what it portends for the shergottite meteorite association. Remember that shergottites and their relatives have crystallization ages of about 1.3 billion years or younger. Comparison of these ages with the duration of igneous activity for various solar system bodies, as shown in Figure 5.11, suggests that these meteorites must be derived from a large planet. However, the problems associated with ejecting rocks from the surface of a planetary body are formidable.

One possible way out of this dilemma is to ascribe the origin of shergottite magmas to impact melting on an asteroidal body. A small amount of melting of target rocks is expected in large impact events. However, this is not at all a satisfying solution. The absence of unmelted rock clasts and the presence of cumulate textures in these meteorites make them very different from the impact melt rocks found in terrestrial and lunar craters. Differences in the initial strontium isotopic compositions of individual shergottites are also inconsistent with this hypothesis, because impact melt sheets appear to be fairly well homogenized.

The evolutionary patterns of the shergottite parent body gleaned from study of these meteorites are also more consistent with a planetary source than an asteroidal source. The isotopic data for shergottites discussed in Chapter 4 suggested that a complex sequence of igneous events was necessary to produce these rocks. Initial differentiation occurred about 4.5 billion years ago, at which time a non-chondritic mantle may have formed. Actually, several mantle source regions with distinct strontium isotopic compositions may have been required to produce the shergottite parent magmas. Partial melting of these source regions followed at approximately 1.3 billion years ago or possibly more recently, and the resulting parent magmas experienced fractional crystallization before or during emplacement. What emerges is a picture of a geologically active body, compositionally similar to the earth and capable of several episodes of igneous activity that spanned much of geologic time. Such complex magmatic behavior is very different from that thought to be possible on asteroids.

The separation of light and heavy rare earth elements in shergottites is so extreme as to require several igneous events to accomplish it. This lends credence to the complex evolutionary pattern already inferred from isotopes. Very few minerals are capable of discriminating between light and heavy rare earths; most minerals just reject them all. One notable exception is the mineral garnet, which has a phenomenal preference for heavy rare earths and is intolerent of the lighter ones. For example, the concentration of ytterbium, one of the heaviest rare earths, in garnets is typically at least 100 times that of lanthanum, the lightest rare earth, relative to their cosmic abundances. One explanation proferred for the highly skewed rare earth pattern measured for shergottites is that garnet was a mineral constituent of the source region. The rare earths would not be distributed equally among the various mantle minerals in this case, and partial melting would release different amounts of light and heavy elements. If this explanation is correct, it also bears on the question of the size of the eucrite parent body. Garnet, an important mineral in terrestrial mantle rocks, is a dense mineral that forms in peridotites only at high pressures. The internal pressures in small bodies are so low that garnet is not stable even at the centers of the largest asteroids. This argument is not as convincing as it might seem, because some other minerals like calcium-rich pyroxenes can distinguish between light and heavy rare earths to a lesser degree and might possibly produce a similar pattern.

Another possible constraint on the size of the shergottite parent body is imposed by the cumulate textures in these meteorites. Computer modeling of the settling of pyroxene grains within a shergottite magma chamber suggests that crystals of this size could accumulate only under the influence of a large gravity field, probably greater than that of the moon. This conclusion is highly dependent on a number of assumptions, but it is one more observation that suggests a planetary parent body.

THE RED PLANET

There is considerable evidence suggesting that the shergottite parent body is a planet. The difficulty in removing rock samples from a planetary gravity field demands that we give preference to the smallest suitable planet we can find. The 1.3–billion-year or younger ages of shergottites appear to limit our consideration to Mars and Venus, so Mars, being the smaller, is the obvious choice. However, the idea that shergottites could be Martian rocks initially met with general disbelief, if not downright hostility. This understandable response stemmed largely from the fact that no lunar meteorites were then known. The logic went something like this: If meteorites could not be ejected from the moon, which is smaller and has a lesser gravitational field than Mars, how could they escape from a larger body? The recent discovery in Antarctica of lunar achondrites has squelched this particular argument, but the exact mechanism by which rocks could have been ejected from Mars is still not understood.

Mariner and Viking photographs of the Martian surface in the 1970s disclosed the existence of gigantic volcanoes. Most of these, like the structures pictured in Figure 5.13, are quite young and

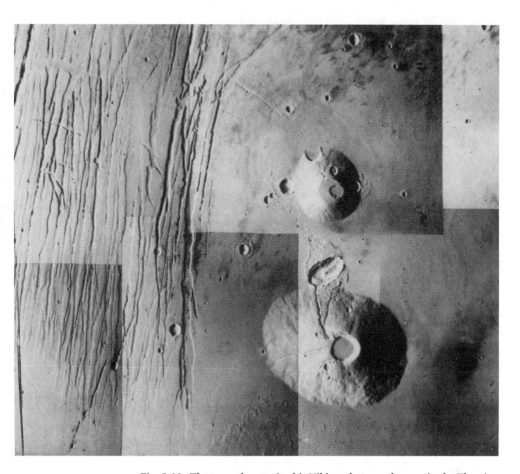

Fig. 5.13. The two volcanoes in this Viking photograph occur in the Tharsis region, the most prominent center of recent igneous activity on Mars. Uranium Tholus (top) is 60 kilometers in diameter, and Ceranius Tholus (bottom) is 120 kilometers across. A relatively recent lava flow has issued from a vent near the summit of the larger volcano. The fractures at the left are part of a set of stress features that were produced by bulging in this region, possibly due to ascending magmas. Photograph courtesy of NASA.

might be suitable sites for shergottite magmatism. Crater densities on volcanic surfaces clearly indicate that Mars was volcanically active during the required period. Are there any other clues that might tie the shergottites specifically to this planet?

When the two Viking spacecraft landed on Mars, they scooped up small shovelfuls of soil and made quantitative chemical analyses. These samples provide an intriguing comparison with shergottites. At both landing sites, the surficial layer of dirt was compacted into "duricrust" and was found to contain high quantities of chlorine and sulfur. This enrichment in volatile elements has generally been interpreted to mean that the uppermost regolith is not simply powdered igneous rock, but has had chlorine and sulfur added from volcanic exhalations. The duricrust-free samples taken from below this altered layer were lower in sulfur and chlorine and presumably had compositions much closer to those of the rocks from which they were formed. The average chemical composition of this duricrust-free soil was very similar to the analyzed compositions of these meteorites. Martian soil was apparently derived from basaltic rocks very similar to these meteorites. This composition is not like those of other known kinds of achondrites.

If shergottites were blasted by impact off the Martian surface, it seems likely that they would have been shocked in the process. Indeed, these are among the most severely shocked meteorites. Several shergottites even contain tiny pockets and veins of impact-melted glass, the ultimate response of rocks to shock. These little glass wedges have provided the most tantalizing and definitive link between the shergottites and Mars. This interesting discovery unfolded almost by accident. An experiment was designed to analyze gases released by heating glass fragments in one Antarctic shergottite, Elephant Moraine A79001. One of the isotopes of gaseous argon, ^{40}Ar, is the decay product of a potassium radionuclide, ^{40}K, and thus forms the basis of a radiometric dating technique. From this experiment, scientists expected to determine the timing of the shock event during which the melt glasses formed, but the age they obtained (in excess of 6 billion years!) was absurd. They correctly reasoned that this glass must contain some extra ^{40}Ar, unrelated to the decay of ^{40}K in this sample. This gaseous isotope was trapped at the time of shock melting, presumably from the atmosphere on the shergottite parent body. The measured abundances and isotopic compositions of argon, krypton, xenon, and nitrogen in the trapped-gas component were subsequently determined. What makes this fascinating is that the trapped-gas component of the meteorite closely matches the composition of the Martian atmosphere measured by Viking landers. The nitrogen measurement is particularly diagnostic because the nitrogen isotopic composition of the Martian atmosphere is so bizarre. Some of the data for isotopes of argon (Ar), krypton (Kr), and xenon (Xe) are shown in Figure 5.14. The values of these isotopic ratios in shergottite melt glasses fall between values for the Martian and terrestrial atmospheres. Even though some contamination by the

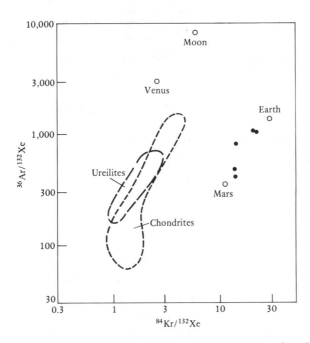

Fig. 5.14. Atmospheric gases trapped in shergottites when they were shock-metamorphosed provide the most compelling evidence that these meteorites may be from Mars. Analyses of the isotopes of gaseous argon (Ar), krypton (Kr), and xenon (Xe) in shock-melted glass of the Elephant Moraine A79001 (Antarctica) shergottite are shown as filled circles. These data points fall between the composition of the earth's atmosphere and that measured on Mars by Viking. These compositions have been interpreted to represent a trapped Martian atmospheric component subsequently contaminated by the terrestrial atmosphere during residence on the earth. These gases are totally unlike those for other solar system materials, also illustrated in this diagram.

terrestrial atmosphere may have occurred, the apparent Martian component is still recognizable.

Two more observations are at least consistent with what we know about Mars, though they cannot be used as evidence for a Martian origin. In addition to volcanic features, large sinuous river channels have been observed on Mars. These are now dry, but their existence indicates that Mars has not always been devoid of water. Thus, hydrous minerals like those found in shergottites would probably be expected in Martian volcanic rocks. The second observation concerns magnetism. The origin of planetary magnetic fields is not fully understood, but they probably result from electric currents generated by fluid motions in metal cores. When rocks cool through a certain temperature in the presence of a magnetic field,

Fig. 5.15. This view of the Martian surface shows vesicular volcanic rocks littering the ground. Could these be shergottites? The metallic object is Viking's soil-collector arm. Photograph courtesy of NASA.

the rocks themselves become magnetized. This occurs because small domains within iron-bearing minerals align themselves with the prevailing global magnetic field. Superimposed on this record of an ancient magnetic field (**paleomagnetism**) is another temporary magnetism induced by the present magnetic field. When a correction is made for the magnetism induced by the earth's field in shergottites, there is not much magnetism left. Put more precisely, shergottites have very low paleomagnetic intensities. These meteorites were heated substantially during shock metamorphism, so the 180-million-year impact event probably marks the time at which shergottites were last magnetized. The weak paleomagnetism therefore suggests that the shergottite parent body had a weak-to-nonexistent magnetic field 180 million years ago. This is consistent with what is known about the present Martian magnetic field, but not with other meteorite parent bodies.

None of these lines of evidence, considered alone, provides an ironclad case that Mars is the shergottite parent body. However, when all of these jigsaw pieces of mostly permissive evidence are fitted together, the case becomes more persuasive. This is often the way science works. Particularly in the geologic and planetary sciences, seldom, if ever, is anything proved in absolute terms, but the accumulation of more and more observations that are consistent with a hypothesis ultimately leads to the consensus that it is correct. This particular consensus is not universally held, however, because of the serious (some say insurmountable) problems in removing rocks of suitable size from the Martian surface. These will be considered in a later chapter.

AN INSCRUTABLE ASTEROID

Examination of the properties of ureilites has unfortunately not led to an understanding of their origin. Aside from the fact that they are igneous cumulate rocks that have been highly shocked, not much else is really clear. This frustrating uncertainty about the formation of ureilites is matched by our ignorance of the ureilite parent body. Discussing this blemish on our heretofore good scorecard serves to emphasize the fact that there are many puzzles left to solve in meteoritics.

Let us make the assumption that the ureilite parent body is an asteroid. This assumption is probably justified by the odds that most meteorites will be derived from small bodies rather than large planets. Spectrophotometric study of ureilites should provide some clues as to the spectral properties of possible parent asteroids. Among this class of achondrites, only the Novo Urei (Russia) meteorite spectrum has been measured. Because of its high carbon content, this sample is very dark, with an albedo of only about 7 percent. Its parent body would thus fall into the C category of asteroids, along with carbonaceous chondrites. Could a ureilite parent body be distinguished from those of carbonaceous chondrites by its spectrum?

The Novo Urei spectral-reflectance curve is shown in Figure 5.16. The spectra of Cl, C2, and C3 chondrites are illustrated for comparison. Also shown is the spectrum for "black" chondrites, ordinary chondrites that have been heavily shocked, resulting in significant color change and lowering of their albedos. Novo Urei seems to be sufficiently different from carbonaceous chondrites to be recognizable by its spectrum, but the severely shocked ordinary chon-

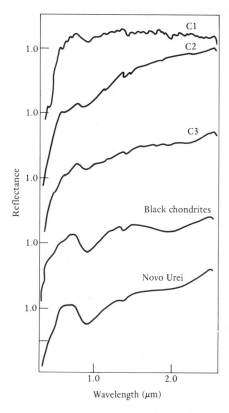

Fig. 5.16. Very little information exists from which to speculate about the ureilite parent body. In this figure, the reflectance spectrum of the Novo Urei ureilite is compared with spectra for C1, C2, and C3 carbonaceous chondrites and heavily shocked ("black") ordinary chondrites. Only the latter are difficult to distinguish from ureilites. Therefore, it seems possible that a ureilite asteroidal parent body might be recognized from its spectra, but none has emerged so far.

drites present a potential problem. There is also some doubt about how representative the Novo Urei spectrum is for ureilites, because this meteorite has been so severely shock-metamorphosed. The spectra of unshocked ureilites would probably have higher albedos. Because the ureilites vary significantly in their shock histories, we do not know which of these spectra, shocked or unshocked, is more appropriate for the ureilite parent body, confounding the problem of identifying it. C-type asteroids make up about three-quarters of all asteroids, so an additional concern is the large number of bodies that must be surveyed. So far, no ureilite parent body has been recognized, but spectrophotometry offers some possibility of success.

MELTED CLUES

A medical practitioner must use all the indirect information that can be gleaned from symptoms and relatively harmless tests to diagnose a patient's illness. There may be more straightforward ways to find out what is wrong, but these might endanger the patient's life. Likewise, for students of the earth's interior, no indirect information can be wasted, because the area under scrutiny is not directly accessible. Almost any kind of natural or man-made vibration must be analyzed for its trajectory and travel time to yield data on interior composition and heterogeneity. Natural force fields must be measured and modeled. Occasional nodules of mantle and deep crustal rocks accidentally transported to the surface by ascending lavas must be chemically and petrologically dissected. Finally, the information from magmas derived from these regions must be extracted.

In the case of achondrite parent bodies, this last bit of indirect evidence is generally the only kind available. These meteorites themselves are the major sources of information on the nature of their parent bodies. There are no direct observations of the orbits of achondrites to tell us from where they have come. Instead, this information must be pieced together from spectral studies, direct comparison with lunar samples, analysis of trapped atmospheric gases, and other, more indirect lines of reasoning.

Among these meteorites we have pieces of the moon, possibly Mars, the asteroid 4 Vesta, and probably a small number of other asteroidal bodies. We have unraveled the records in these rocks concerning the composition and depth of source regions, the timing of igneous events, and the internal stratification of parent bodies. These data would be useful even if we had no idea of where achondrites came from, but their importance is magnified when parent bodies can be identified.

SUGGESTED READINGS

There are several excellent nontechnical publications on the moon, but almost nothing on other achondrite parent bodies. The other references cited here are intended for the reader who is interested in more detail, and they may require some additional scientific vocabulary.

GENERAL

McSween H. Y. Jr. and Stolper E. (1980) Basaltic meteorites. *Scientific American* 242, 54−63. (Nontechnical paper describing the eucrite and shergottite associations and their possible parent bodies in easily understood terms.)

ASTEROIDAL SOURCES

Drake M. J. (1979) Geochemical evolution of the eucrite parent body: Possible nature and evolution of asteroid 4 Vesta? In *Asteroids*, edited by T. Gehrels, University of Arizona Press, Tucson, pp. 765−782. (Technical paper that cogently presents the case for Vesta as the eucrite parent body.)
Miyamoto M. and Takeda H. (1977) Evaluation of a crust model of eucrites from the width of exsolved pyroxene. *Geochemical Journal* 11, 161−169. (Technical paper in which an internal model for the eucrite parent body is constructed from mineralogical data.)

PLANETARY SOURCES

Wood C. A. and Ashwal L. D. (1981) SNC meteorites: Igneous rocks from Mars? In *Proceedings of the 12th Lunar and Planetary Science Conference*, Pergamon Press, New York, pp. 1359−1375. (Technical paper presenting the evidence for a Martian origin for shergottites.)
McSween H. Y. Jr. (1984) SNC meteorites: Are they Martian rocks? *Geology* 12, 3−6. (A brief but up-to-date technical review summarizing the arguments for a Martian origin for shergottites.)

THE MOON

Taylor S. R. (1975) *Lunar Science: A Post-Apollo View*, Pergamon Press, Oxford, England, 372 pp. (An engrossing, generally nontechnical account of lunar science as reshaped by the Apollo program.)
French B. M. (1977) *The Moon Book*, Penguin Books, Baltimore, 287 pp. (A richly illustrated, nontechnical review of lunar science that is not mired in terminology.)

ASTEROIDAL SPECTRA

McCord T. B., Adams J. B., and Johnson T. V. (1970) Asteroid Vesta: Spectral reflectivity and compositional implications. *Science* 168, 1445−1447. (A classic technical reference that first presented evidence for an asteroidal parent body for eucrites.)

Fig. 6.1. One of the most spectacular meteorite finds in Antarctica was the Derrick Peak 78009 iron meteorite. Sixteen specimens like this one were found among angular blocks of sandstone on the slopes of this nunatak. Photograph courtesy of NASA.

6 Iron and stony-iron meteorites

One of the most significant meteorite finds in Antarctica was made accidentally by a New Zealand geologic field party in 1978. While working on the slopes of Derrick Peak, a remote nunatak free of ice and snow, this group stumbled upon a number of iron meteorites. News of their discovery was soon transmitted by radio to a joint American-Japanese meteorite search team at a nearby base camp on the Darwin glacier. Team scientists quickly boarded a helicopter and arrived in time to aid in the search for more specimens. In all, 16 samples were recovered. Several, like the meteorite in Figure 6.1, were quite handsome and of sufficient size to make carrying them down the steep mountainside difficult.

It is appropriate that iron meteorites like those at Derrick Peak would be found in Antarctica, the most inaccessible and inhospitable of places. These chunks of metal were forged in the deep and once-infernal interiors of extraterrestrial bodies, and it somehow seems right that we endure some hardship to get them. This chapter will show that the information carried in iron and stony-iron meteorites is worth all that effort.

THE CORE OF THE PROBLEM

The earth appears to be approximately chondritic in composition. This means that its most important element by weight is iron, composing nearly 40 percent of the whole planet. Aside from a few ore deposits that contain iron concentrated by igneous or sedimentary processes, no common rocks contain anywhere near the chondritic abundance of this element. So where is all this missing iron? It is located in a massive central core of metal, approximately 6,940 kilometers across, that formed during the initial differentiation of the earth.

Ample evidence for this core comes from a variety of geophysical measurements. The mean density (mass/volume) of the earth, about 4.0 to 4.5 grams per cubic centimeter after correction has been made for compression due to overlying materials, is significantly greater than that of common rocks. This indicates that more massive material must be hidden in the planet's interior, and iron metal is certainly denser than any kind of rock. Earthquakes produce seismic vibrations with periods of a few seconds that are transmitted through the earth with velocities dependent on the densities of the rocks through which they pass. Seismic waves reflected off the core permit its size to be determined, and the measured velocities of seismic waves propagated through the core allow an estimate of its composition. Interior disturbances may also produce free oscillations, gross vibrations of the whole planet acting as if it were a ringing bell. These oscillations, with periods ranging from minutes to several days, constrain how materials of different densities are distributed inside the earth. It is well known that the earth is not perfectly spherical, but bulges at the equator. This flattening is due to rotation and is also an expression of the way mass is distributed internally. The moment of inertia, calculated from this rotational flattening, as well as constraints from free oscillations indicate a large concentration of mass at the earth's center.

The earth's core is inaccessible to us for direct study, so we must rely almost completely on these indirect measurements. One writer has noted that reaching conclusions from this kind of evidence is like trying to reconstruct the inside of a piano on the basis of the sounds it makes while crashing down a staircase. However, it is logical to think that core samples would look much like iron meteorites, which likewise resulted from some sort of core-forming processes. Partial melting of chondritic material in asteroids would produce liquid metal that, because of its higher density, would drain away from silicate melt. The commercial smelting of iron depends on this tendency for molten metal and silicate to separate. In the presence of a gravity field, this metallic liquid would tend to displace solid silicates and sink to form a central core, but the complete segregation of metal depends on a number of factors that might not be operable in small bodies.

In discussing achondrites in Chapter 5, we saw that partial melting occurred within some asteroidal parent bodies. The bulk composition of the eucrite parent body was approximately that of chondrites, which normally contain abundant metal, but almost no metal is found in the eucrites themselves. The missing metal may have segregated into a core. There is thus a kind of complementary relationship between achondrites and irons, and it is probably a valid assumption that bodies with differentiated cores also contain achondrites.

METAL-LOVING ELEMENTS

The idea that individual elements have distinctive geochemical behaviors was first appreciated by V. M. Goldschmidt at the University of Oslo in the early part of this century. His pioneering contributions in this area quickly established his scientific reputation, but his work was interrupted when the Germans invaded Norway in 1940. Although Goldschmidt escaped to England in 1942, his health had been so impaired by imprisonment in concentration camps that he died soon thereafter.

In 1923 Goldschmidt had coined the terms **siderophile, chalcophile**, and **lithophile** to describe elements with tendencies to concentrate in metal, sulfide, and silicate, respectively. Although this has proved to be a serviceable geochemical classification, not all elements can be neatly pigeonholed, because their affinities may vary with temperature, pressure, or chemical environment. When this concept of geochemical affinities was first proposed, there were few quantitative data with which to test the idea. It is probably obvious that the geochemical behavior of an element can be determined from experiments that measure its partitioning between coexisting metal, sulfide, and silicate phases, but carrying out such experiments remains a difficult task even today. Goldschmidt had the insight to recognize that meteorites represent fossilized natural experiments of this kind. He and his coworkers spent years separating and analyzing numerous meteorite components in order to predict geochemical affinities. The conclusions of this work have been confirmed by studies of smelting products that contain a silicate slag, sulfide matte, and metallic iron.

Core formation, whether in planets or in planetesimals, has the effect of scavenging siderophile (metal-loving) elements. Elements exhibiting the most rabid siderophile tendencies are iron (Fe), nickel (Ni), cobalt (Co), platinum (Pt), osmium (Os), iridium (Ir), gold (Au), palladium (Pd), ruthenium (Ru), and rhodium (Rh). Of these elements, only iron and nickel have high enough cosmic abundances to make them volumetrically important, so iron meteorites are composed mostly of Fe-Ni alloys. The range of nickel in iron meteorites is illustrated in Figure 6.2. No irons contain less than 5 weight percent nickel, and only a handful contain more than 20

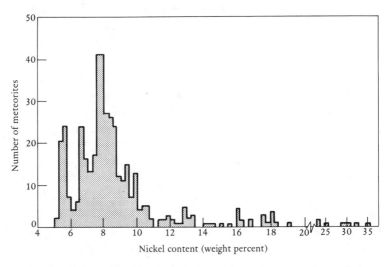

Fig. 6.2. Of the siderophile elements that make up iron meteorites, only iron and nickel have high enough cosmic abundances to occur in major proportions. This diagram summarizes the compositions of hundreds of analyzed iron meteorites. Nickel contents are highly variable, but most meteorites contain between 6 and 10 percent of this element. No iron meteorites contain less than 5 percent nickel, and only seven contain more than 20 percent.

percent. Several other elements that exhibit siderophile behavior under the right conditions are germanium (Ge), gallium (Ga), phosphorus (P), and carbon (C). These occur in only minor quantities in iron meteorites, but some are important for classification purposes.

The compositions of iron meteorites are not restricted only to siderophile elements, although these compose the great bulk of such meteorites. Experiments have disclosed that a mixture of Fe-Ni metal with sulfur has a lower melting point than metal alone. If iron meteorites formed by partial melting, it is to be expected that they should contain some sulfur, as indeed they do. Where sulfur goes, chalcophile (sulfide-loving) elements follow. However, iron meteorites contain less sulfur than the amount necessary to make this low-melting-point mixture. The reason for this apparent sulfur depletion is not known.

In Chapter 1, we discussed the hypothesis that a large meteoroid impact caused the extinction of dinosaurs and other organisms, marking the end of the Cretaceous Period. The evidence presented for this idea was that high concentrations of iridium and gold have been measured in sediments deposited during that time, and these same elements are also concentrated in meteorites. Now that we have seen how elements behave in geochemical systems, we are in a better position to evaluate this evidence. The anomalous elements concentrated at the Cretaceous-Tertiary boundary are all siderophile in character. Chondrites, the most common type of meteoroid impacting the earth, are undifferentiated and thus have their full cosmic complement of siderophile elements. Therefore, such meteorites could be expected to add high concentrations of iridium and gold relative to terrestrial crustal rocks, which were already depleted in siderophile elements by core formation.

ASSEMBLY DIRECTIONS FOR IRONS

The assembly of siderophile and, to a lesser extent, chalcophile elements to form iron meteorites is a more complex task than it might first appear. Over 40 different minerals, a number of which are unknown in terrestrial rocks, have been identified in iron meteorites; however, most of these are present in very minor quantities. Only a few of the most important minerals will be considered here.

Iron-nickel alloys, which compose the great bulk of iron meteorites, are of two basic types: Kamacite contains up to 7.5 weight percent nickel, and taenite varies in its nickel content between about 20 and 50 percent. Although both minerals have crystal structures with cubic symmetry, the different sizes of nickel and iron atoms cause these minerals to have different architectural styles, as illustrated in Figure 6.3. Kamacite forms a "body-centered" lattice, with each atom located at the center of a cube, so that it is surrounded by 8 neighboring atoms. The "face-centered" lattice of taenite features an atom centered on each face of a cube, so that every atom in this structure has 12 neighbors. It should be obvious that the face-centered lattice is a more efficient way of packing atoms within a given volume. Kamacite and taenite are important components of steel and are known to metallurgists as alpha iron and gamma iron, respectively.

Masses of liquid metal solidified to form iron meteorites as they cooled below roughly 1,400°C. However, the crystallization histories of these metal chunks did not stop there, but continued as solid-state recrystallization occurred at lower temperatures. Shown in Figure 6.4 is the **phase diagram** for the Fe-Ni system below 1,000°C. Phase diagrams are useful for showing the fields of stability of various minerals in terms of temperature and composition.

Crystal structures for metals

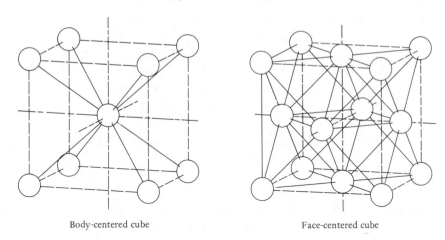

Body-centered cube   Face-centered cube

Fig. 6.3. Iron-nickel alloys, from which iron meteorites are formed, exhibit two basic crystal structures. Kamacite forms a body-centered lattice, with each atom located at the center of a cube so that it is surrounded by 8 other atoms. Taenite occurs as face-centered cubes, with an atom centered on each face. In this structure, each atom is surrounded by 12 neighboring atoms.

Above 900°C, metal of any composition has the taenite structure, but at lower temperatures low-nickel kamacite and high-nickel taenite begin to unmix in the solid state. This unmixing takes place when metal cools to the boundary of the kamacite + taenite field on the phase diagram. From the slope of this boundary, it is apparent that the temperature at which this happens depends on the nickel content of the metal. Kamacite continues to form down to about 500°C, at which point migration of atoms through the solid metal becomes so sluggish that unmixing stops.

In order to see how iron meteorites of different compositions crystallize, let us take three specific examples, illustrated by arrows on the Fe-Ni phase diagram. On cooling, an alloy containing 5 percent nickel will begin to change from the taenite structure to the kamacite structure at about 800°C. Below about 650°C this metal will enter the kamacite field, and the transformation will be complete. The final product will consist only of kamacite. At the other end of the compositional scale, an alloy with 30 percent nickel will just about reach the temperature limit for atom migration (500°C) before any unmixing starts. This meteorite, when finally cooled, will consist almost entirely of taenite. The most interesting

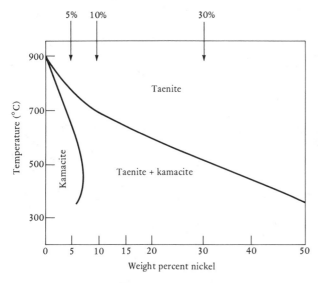

Fig. 6.4. A phase diagram predicts which phases should be stable at various combinations of temperature and chemical composition. The phase diagram for the iron-nickel system below 1,000°C, constructed from a series of experiments by metallurgists, is divided into three fields showing the stability of taenite, kamacite, or both. Cooling of an iron meteorite with a certain nickel composition would be represented on this diagram by dropping a vertical line from the top. A mass of solid metal will change its structure as it moves into a different stability field during cooling.

situation is something in between these two extremes, as illustrated by a l0 percent nickel alloy. It also turns out that intermediate nickel values are the most common in iron meteorite compositions. In this case, the taenite cools to about 700°C and begins to unmix. By 500°C, this composition resides within the field where taenite and kamacite coexist, so the final product will contain both minerals.

The physical appearance of iron meteorites containing both kamacite and taenite is striking. Grains of kamacite nucleate and grow in certain preferred orientations within the face-centered taenite lattice. Kamacite forms four sets of plates that cut the corners of the original cubic crystals, forming eight-sided octahedra. The intergrown plates of kamacite within taenite can be readily seen in slices of iron meteorites that have been polished and etched with dilute acid. In l808, Count Alois de Widmanstatten, director of the Imperial Porcelain Works in Vienna, first observed this structure, now called the **Widmanstatten pattern**. The geometry of the

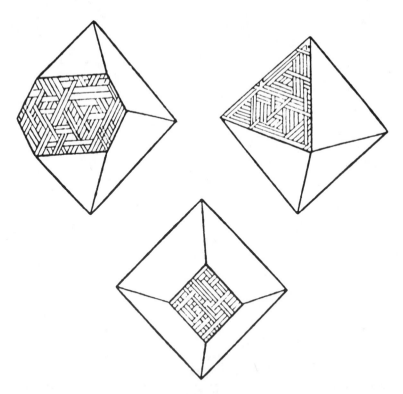

Fig. 6.5. The Widmanstatten pattern is a regular geometrical intergrowth of kamacite and taenite that forms during slow cooling of iron meteorites of appropriate composition. These sketches of the Widmanstatten pattern as it would appear in various cuts through the original taenite crystal were made by G. Tschermak in 1894.

intergrown kamacite plates, of course, varies depending on the orientation of the meteorite when it was sawed. Sketches of how the Widmanstatten pattern would appear in various sections through a parent taenite crystal are reproduced in Figure 6.5.

Randomly interspersed among the metal grains in most iron meteorites are other minerals also rich in iron and/or nickel. Troilite (FeS) is almost ubiquitous, forming dark, rounded nodules. Inclusions of this phase were first observed by the Jesuit priest Domenico Troili in 1766, but a century elapsed before accurate chemical analyses proved that it was distinct from pyrite (FeS_2), the "fool's gold" commonly found in terrestrial rocks. Sulfur is highly soluble in molten iron, but practically insoluble in solid iron, so it is necessary for this element to form a sulfide as the metal solidifies.

Carbon and phosphorus are also soluble in metallic liquids, and small amounts of these elements (generally less than 0.1 percent) may substitute in iron alloys. But, as with sulfur, the bulk of carbon and the bulk of phosphorus end up in their own phases. Carbon forms cohenite, $(Fe,Ni,Co)_3C$. This mineral is unstable at all temperatures under conditions of very low pressure and breaks down into metal (kamacite) plus pure carbon (graphite). That cohenite occurs at all is due to the fact that its decomposition is extremely slow, and different meteorites show this breakdown in various stages of arrest. Phosphorus forms the mineral schreibersite, $(Fe,Ni)_3P$. This is a particularly attractive mineral, commonly occurring as tiny white prisms within metal grains. Any chromium in the original melt forms chromite, $FeCr_2O_4$. Most of the other elements in iron meteorites are present in only trace amounts and display enough siderophile behavior to form solid solutions with metal.

ORDER OUT OF CHAOS

Upon casual inspection, iron meteorites seem to be highly diverse and perplexing samples, but order is restored by proper classification. A good classification system should do more than simply superficially lump similar meteorites together. Ideally it should provide information on their origins. There are two entirely different classification schemes for iron meteorites now in use. One of these is based on the structures present in irons, and classification can be done by visual inspection. The other requires accurate and time-consuming chemical analyses. Both systems produce similar groupings in most cases, but the chemical classification provides more genetic constraints. We shall first consider the **structural classification**.

Irons with less than about 6 weight percent nickel contain kamacite but no taenite. The sizes of the original kamacite grains must have been larger than the present dimensions of the meteorites, for in most cases these irons are single crystals. Because kamacite is cubic, such meteorites are called **hexahedrites** (abbreviated H), from the Greek word for "cube." Polished surfaces of these irons are featureless, except for numerous **striations** in some meteorites. An example of a hexahedrite with such striations, called **Neumann lines**, is shown in Figure 6.6. These lines are boundaries between twins (segments of the same crystal in different orientations) formed by shock deformation.

Octahedrites range in nickel content between approximately 6 and 17 weight percent. These meteorites contain both kamacite and taenite in the decorative Widmanstatten pattern. The octahed-

Fig. 6.6. The Calico Rock (Arkansas) hexahedrite consists almost entirely of kamacite. This cut face shows Neumann lines formed by shock deformation, but is otherwise featureless. The meteorite measures approximately 8 centimeters across. Photograph courtesy of the Smithsonian Institution.

rites constitute by far the largest and most diverse group of irons, so it is advantageous to subdivide them. A convenient way of doing this is by making use of the widths of the kamacite bands that form the Widmanstatten pattern. It is conventional to use a fivefold subdivision of octahedrite band widths, each formed from the previous smaller one by multiplication times 2.5. In this system, octahedrites are divided into the following groups: coarsest (Ogg), coarse (Og), medium (Om), fine (Of), and finest (Off). Examples of some structural differences among octahedrites are shown in Figures 6.7 and 6.8.

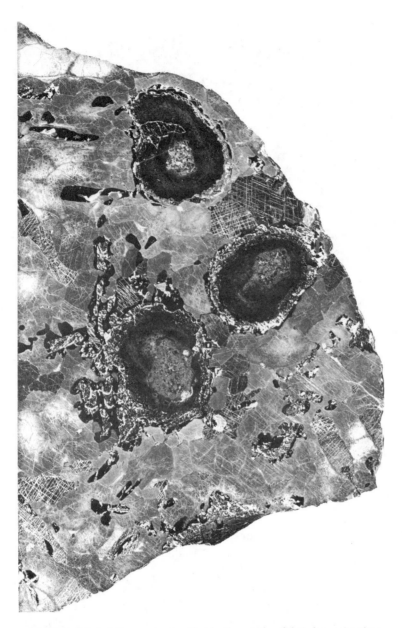

Fig. 6.7. Octahedrites can be classified by the widths of their kamacite plates. This slab of the Mount Stirling (Australia) coarse octahedrite exhibits a coarse Widmanstatten pattern. The large black ovals are inclusions of the sulfide mineral troilite. The bottom cut surface of the meteorite measures about 3 centimeters. Photograph courtesy of the Smithsonian Institution.

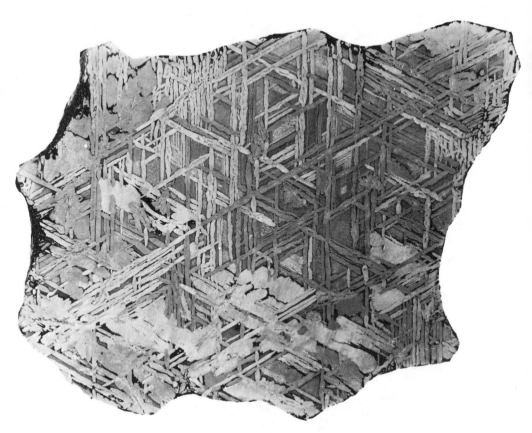

Fig. 6.8. The Waingaromia (New Zealand) fine octahedrite contains thin, oriented bands of kamacite that form its Widmanstatten pattern. The width of this slab is approximately 11 centimeters. Photograph courtesy of the Smithsonian Institution.

Finest octahedrites actually grade into meteorites with no obvious structure. These irons, which have high nickel contents, are called **ataxites** (D). The name is derived from a Greek word meaning "without order." This is perhaps an unfortunate term, because ataxites do have a microscopic Widmanstatten pattern, but it is not perceptible to the naked eye. Such meteorites consist almost entirely of taenite, with only a few microscopic plates of kamacite. An ataxite is pictured in Figure 6.9.

A number of iron meteorites cannot be conveniently assigned to one of these structural classes. These are commonly rather fine-grained, consisting of centimeter-sized crystals rather than the usual larger grains ranging up to meter-sized. Iron meteorites that resist classification are quite logically called **anomalous** (Anom).

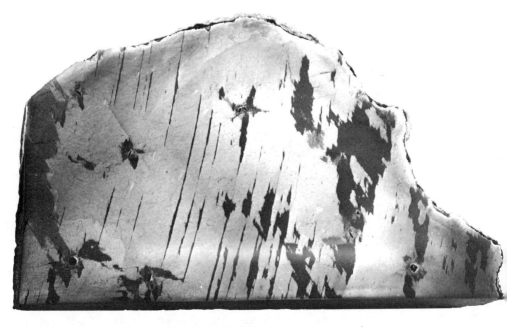

*Fig. 6.9. Ataxites have no readily observable Widmanstatten structure, be-
cause they contain only minute quantities of kamacite. Hoba (Namibia), the
largest recovered meteorite, is of this type. This slab of the Hoba ataxite is featureless except for the dark bands that were produced by shock. The bottom cut
face measures about 15 centimeters. Photograph courtesy of the Smithsonian Institution.*

The **chemical group** classification system for irons utilizes trace
elements to diagnose the different groups. An initially proposed
subdivision into four groups, named I through IV, was based on
varying contents of gallium and germanium as well as the abundance of nickel. When enough high-quality chemical data became
available, some of these groups were subdivided even further. For
example, group IV has now been separated into IVA and IVB, which
apparently have nothing to do with each other. To make things
even more complicated, a few subdivided groups have now been
recombined, after transitional members were discovered. This had
led to the somewhat confusing situation of having some group
names like IIAB. The resulting motley array of symbols is very
difficult to remember, but otherwise this is a very serviceable classification system. Most of the approximately 600 known iron meteorites have now been analyzed. From this large data base it is
possible to recognize twelve important groups with five or more
members each. When the measured concentrations of germanium

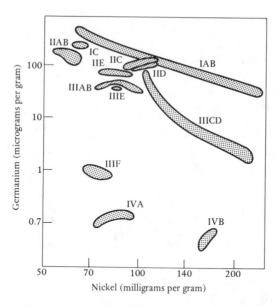

Fig. 6.10. *A logarithmic plot of the measured concentrations of germanium versus nickel in iron meteorites provides a way to classify them into different chemical groups. The 12 groups shown account for 86 percent of all analyzed irons, the others being anomalous. Other elements besides germanium and nickel can also be used to classify these meteorites.*

and gallium are plotted against nickel on logarithmic scales, these subdivisions are clearly resolved. Figure 6.10 illustrates the clusters representing these twelve groups on a germanium versus nickel abundance plot. A graph of gallium versus nickel looks rather similar. Meteorites that do not plot within these well-defined chemical clusters are again called anomalous. Keep in mind that anomalous irons in the chemical sense may not be anomalous in the structural sense.

The meteorites in each chemical group generally belong to a limited range of structural classes, as shown in Table 6.1. This is actually somewhat surprising, because the structural classification is not solely dependent on composition. The kamacite band widths on which this system is based are dictated not only by the nickel content of the cooling taenite crystals but also by the rate at which these crystals cooled. The metallurgist who proposed the structural classification for iron meteorites actually cheated a little bit in defining his band-width intervals so that they would most nearly coincide with chemical groups. However, this definition of structural classes is fully justified by the fact that there is a correlation

Table 6.1. *Comparison of chemical and structural classifications for iron meteorites*

Chemical group	H	Ogg	Og	Om	Of	Off	D	Anom	Proportion (%)
IAB			———	——					18.7
IC			——					——	2.1
IIAB	———	——							10.8
IIC						——			1.4
IID				———	——				2.7
IIE			——	——				——	2.5
IIIAB				———	——				32.3
IIICD					———	——	——		2.4
IIIE			——						1.7
IIIF		———	——	——	——				1.0
IVA					——				8.3
IVB								——	2.3
Anom	———	——	——	——	——	——	——	——	13.8

in most cases between band width and chemical grouping. The few iron meteorites for which the correlation does not hold can be rationalized as the products of cooling rates different from those for the rest of their chemical group.

SOLIDIFICATION OF CORES

Chondrites, which have never been melted, contain metal that is much more uniform in composition than metal in iron meteorites. For example, the total range of iridium abundances in several groups of irons varies a thousandfold, whereas iridium concentrations in chondrite metal vary over less than a factor of two. These large variations within iron meteorite groups are attributable to the crystallization processes by which cores were formed. Fortunately, these variations are not random. Within a given chemical group, irons can be ranked in a sequence such that some elements (nickel, gold, arsenic, cobalt, palladium, phosphorus, and molybdenum) increase while others (iridium, osmium, platinum, ruthenium, tungsten, and chromium) decrease. When plotted on logarithmic scales, the concentrations of such elements in meteorites of any one classified group form straight lines. One example of such trends is shown in Figure 6.10. This consistent pattern from core to core is not accidental, but arose through fractional crystallization as

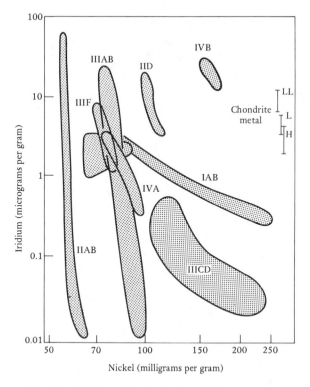

Fig. 6.11. *This logarithmic plot shows iridium versus nickel contents in iron meteorite groups. The iridium abundances extend over many orders of magnitude because of processes occurring during solidification of parent-body cores. These large ranges of iridium abundances are very different from the unfractionated compositions of metals in H, L, and LL ordinary chondrites. The parallel, nearly vertical trends for each iron group, except IAB and IIICD, were caused by fractional crystallization.*

melted metal separated from solid metal. **Partition coefficients** (the ratios of element concentrations in the solid to those in liquid metal) have been measured in the laboratory, and these correctly predict the trends of these elements observed within iron groups. Nickel and other elements with partition coefficients less than 1 are concentrated preferentially in liquid, whereas elements like iridium with partition coefficients greater than 1 accumulate in the solid metal. Fractional crystallization that removes liquid from solid produces linear trends with negative slopes on the iridium versus nickel diagram shown in Figure 6.11.

A few element trends are not linear. For example, germanium in group IIIAB irons forms a boomerang-shaped array when plotted versus nickel, as shown in Figure 6.10. This occurs because some partition coefficients vary depending on what other elements are present. For example, phosphorus has an effect on the distribution of germanium between solid and liquid iron. Germanium is apparently somewhat of an opportunist; it favors liquid iron when phosphorus contents are low, but switches its preference to solid metal at high phosphorus contents. During fractional crystallization of the IIIAB core, phosphorus contents increased, so the slope of the germanium-nickel trend changed.

Two groups of iron meteorites, IAB and IIICD, are unusual in that their fractionation trends are different from those of other groups. In the iridium-versus-nickel diagram (Figure 6.11), the slopes of these two trends are much shallower than those of the other groups. Such trends appear to violate the rules for the igneous behavior of these elements determined from experiments. The inference is that these two iron groups did not crystallize from complete melts. However, trace-element patterns can be modeled successfully by assuming that they represent mixtures of metallic melts with unmelted iron. The implication of this observation is that these two individual iron groups may represent an arrested early stage of core formation. Thus, we can study cores at various points in their evolutionary history by examining irons from different parent bodies.

The effects of fractional crystallization or incomplete melting in cores have been to smear the chemical compositions of individual iron meteorites belonging to any particular group. However, much can still be learned about the original core compositions, because we know from experiments in which directions these processes would alter them. Overall, we gain from these added complications because intermediate steps in the partial melting and crystallization processes that affected cores can be studied and understood.

COOLING INFERNOS

The reason that structural classes of iron meteorites vary with cooling rate is because the Widmanstatten patterns form by unmixing in the solid state, which is a temperature-controlled process. The longer iron meteorites are held at high temperatures in cores within their parent bodies, the more unmixing will occur. Quantitative measurement of these cooling rates could provide important constraints on the sizes of the parent bodies in which irons formed.

How has the record of cooling rate been fixed in these meteorites, and how can we play back this record to obtain such information?

In order to understand this, we must return to the Fe-Ni phase diagram. Again consider a mass of metal containing 10 percent nickel, which will crystallize from liquid into taenite and cool along a vertical line in Figure 6.4. At about 700°C, this alloy will begin unmixing to form oriented bands of kamacite, the composition of which is about 4 percent nickel. The boundaries to the kamacite + taenite field give the compositions of the two metal phases coexisting at any one temperature. These sloping field boundaries indicate that the nickel contents of both kamacite and taenite increase with further cooling. For example, when the mass has cooled to 600°C, the kamacite composition is about 6 percent nickel, and that of taenite is about 19 percent nickel. How can this be? The only way that both metal phases can increase their nickel contents while the composition of the whole mass remains constant is to increase the amount of the low-nickel phase (kamacite) at the expense of the high-nickel phase (taenite). The kamacite plates must become thicker, and taenite between them must shrink by an equivalent amount.

The growth or shrinking of metal phases occurs by migration of atoms within the solid metal grains. Nickel atoms at a grain margin creep into the interior by successive exchanges with iron atoms until the whole grain is homogeneous in composition. This process, called **diffusion**, occurs more rapidly at high temperatures than at lower ones. At some point, about 700°C for taenite, diffusion becomes sluggish enough that nickel atoms can no longer travel freely into grain interiors, and they begin to pile up at the margins. This situation is a little like a traffic jam, where constriction of the flow of cars occurs first at one point and then propagates back along the highway. This slowdown of diffusion results in zoned taenite in which the nickel content decreases toward the center of the grains. By about 500°C, diffusion effectively stops, and the zoning profile is frozen in. An example of this is shown in Figure 6.12. An electron-microprobe traverse through a taenite grain from one boundary to the other shows an *M*-shaped nickel profile, with high nickel at the edges and lower nickel in the center.

Widmanstatten patterns formed at cooling rates so slow that they cannot be easily reproduced in the laboratory. Nevertheless, understanding how these patterns formed permits the growth of this pattern to be modeled theoretically. The final nickel distribution

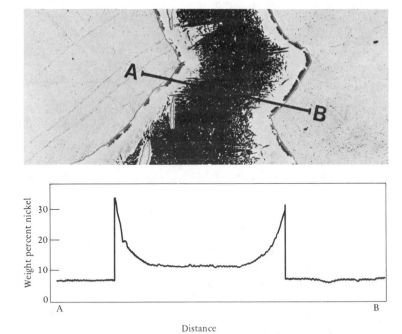

Fig. 6.12. *A microscopic view of the Widmanstatten pattern in the Drum Mountains (Utah) iron meteorite shown above is seen in reflected light. The white taenite area has a dark center of plessite, a very fine intergrowth of taenite and kamacite. An electron-microprobe traverse measuring the nickel content across this region would have the kind of M-shaped profile shown below. The nickel content decreases away from the margins of the taenite grain because diffusion during cooling became too sluggish at low temperatures to homogenize the grain. Such nickel profiles provide a means of calculating the cooling rates of iron meteorites.*

across any taenite grain in an intergrowth with kamacite will depend on the bulk nickel content of the mass and the size of the grains through which diffusion occurred. These can be measured in the laboratory, along with the diffusion rates of nickel at various temperatures. These data have been used to develop computer models for the growth of kamacite bands and the resulting nickel zoning profiles in taenite grains at different cooling rates. The only action that is then required to calculate the cooling rates at which iron meteorites formed is to measure the *M*-shaped profiles in actual taenite grains and compare these with the theoretical profiles.

The cooling rates derived by this method vary from less than one degree to several thousand degrees per million years. These rates are slow, yet they are still appropriate to small asteroidal bodies. The cores of larger planetary bodies like the earth have not yet cooled appreciably in over 4 billion years.

SILICATE INCLUSIONS

The scarcity of silicates in iron meteorites is not surprising, because the radical differences in density between these materials would presumably cause the lighter silicates to float out of metallic cores. However, several classes of irons do contain **silicate inclusions**, and in some cases they are reasonably abundant, as illustrated in Figure 6.13.

These bits and pieces of nonmetallic minerals are founts of information that is not otherwise obtainable for iron meteorites. For example, direct isotopic dating of irons is not possible because they contain almost no suitable radionuclides. The silicate inclusions, however, can be dated by various radiometric methods. Rubidium-strontium isotopes indicate that group IAB silicate inclusions are 4.45 ± 0.1 billion years old, similar to the ages of chondrites. These particular inclusions appear to be highly recrystallized chondritic material, and their ages presumably record the time of their metamorphism. It seems possible that core formation occurred at that same time, although the angular shapes of these inclusions suggest that their relation to their metallic host is that of clasts in a breccia. In contrast, the rounded silicate inclusions in IIE irons have igneous textures and non-chondritic compositions, indicating that they formed as globules of fractionated silicate magma in a mass of cooling metal. Their ages are 4.60 ± 0.1 billion years, except for inclusions in one meteorite that give an age of about 3.8 billion years. If this last age is correct, it suggests that differentiation of the IIE parent body was a protracted process.

We have already seen that oxygen isotopes provide a useful tool for indicating possible relationships between meteorites. These have also been measured in silicate inclusions, and the available data are shown in Figure 6.14. Inclusions from IIE irons define a mass-fractionation line through H chondrites. Group IVA inclusions plot along a similar line through L and LL chondrites. Such links suggest that these iron meteorite groups may represent the cores of differentiated objects that were formerly ordinary chondrite parent bodies. IAB inclusions are different from all the major chondrite groups, but are similar to the anomalous chondrite Winona (Ari-

Fig. 6.13. Some iron meteorites, like the Pitts (Georgia) octahedrite shown here, contain silicate inclusions. The large gray areas with irregular shapes in this photograph are the inclusions. Each of these is 1 to 2 centimeters in diameter. Silicate inclusions are useful sources of information that cannot be obtained from metallic minerals. Photograph courtesy of the Smithsonian Institution.

zona). This meteorite is probably a representative of another sparsely sampled chondrite class.

ADDED COMPLICATIONS

Iron meteorites are already difficult enough to understand, but nature and man in some cases have joined forces to make this task even more taxing. These added complications take the form of new structural modifications superimposed on the original ones.

One cause of these changes is deformation due to shock, either by impacts in space or during arrival on the earth. Iron meteorites are not commonly brecciated as are stony meteorites, but the results of impacts are nevertheless present. One prominent example

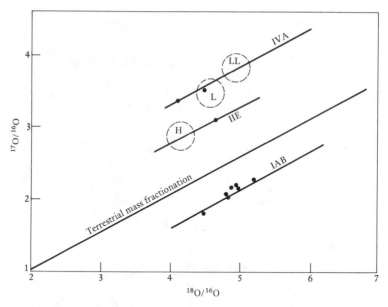

Fig. 6.14. Oxygen isotopes in silicate inclusions of iron meteorites indicate affinities with various groups of chondrites. Analyses of these inclusions are shown as dots in this figure. Inclusions in IAB irons are similar to those in several unique chondrites, those in IIE irons are like H chondrites, and those in group IVA resemble L or LL chondrites. These similarities suggest that the parent bodies for these groups of iron meteorites were chondritic before melting and differentiation to form metallic cores.

is the development of Neumann lines, due to twinning, already illustrated in Figure 6.6.

Another kind of change is induced by heating, which anneals the original structures. Friction during atmospheric passage produces heat, but usually an annealed exterior rind of only 5 to 20 millimeters thickness is created by this rapid transit. Of the 600 or so known irons, 94 have been heated artifically by man in his curiosity and eagerness to test or utilize these metals. A number of iron meteorites have also been used as raw materials to make various artifacts, such as decorative beads or swords. Furthermore, our forefathers showed a propensity to maim iron meteorites by hammering and chiseling them. The Widmanstatten patterns of 8 large irons were distorted by such cold-working when these masses were used as anvils.

Fig. 6.15. Because they are composed of metal, iron meteorites have been used in the past to make artifacts. The Jalandahr knife was forged from a iron meteorite that fell in Punjab (India) in 1621. Photograph courtesy of the Smithsonian Institution.

PALLASITES

From the name **pallasite**, one might logically (but incorrectly) infer that these are samples of the asteroid 2 Pallas. Actually, the only link between the asteroid and the meteorites is that they take their names from the same person, the German natural historian P. S. Pallas. In 1776, Pallas wrote a lengthy description of one of these meteorites. He had earlier been invited by the Russian monarch to explore the vast uncharted region of Siberia, and one of the curiosities he obtained on this expedition was the first known pallasite.

Pallasites are stony-iron meteorites composed almost exclusively of abundant olivine grains enclosed in metal, as illustrated in Figure 6.16. The relative amounts of these two constituents vary, but an olivine-to-metal volume ratio of about 2:1 is common. This is approximately the predicted proportion of olivine where these silicate grains are close-packed spheres of uniform size. Where metal is more abundant, it has a well-developed Widmanstatten pattern.

The composition of olivine in most pallasites is very rich in magnesium, and metals have nickel, germanium, and gallium contents that are rather similar to those in group IIIAB irons (the most populous group with the boomerang-shaped pattern on the germanian-nickel plot). Most pallasite metals actually define an extension of the IIIAB trend. The composition of pallasite metal appears to

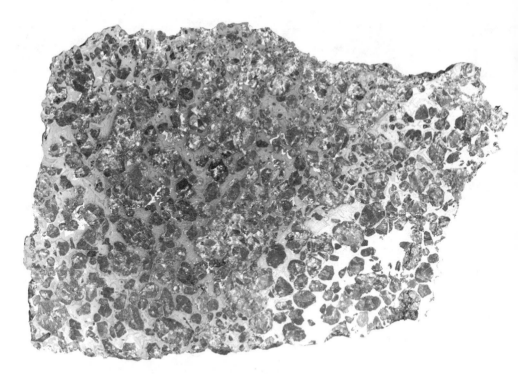

Fig. 6.16. *Pallasites are stony-iron meteorites consisting of olivine grains enclosed by metal. This specimen from Salta (Argentina) measures about 14 centimeters in its short dimension. Photograph courtesy of the Smithsonian Institution.*

be that of the liquid that would be left after extensive crystallization of a IIIAB melt.

A few pallasites, typified by the Eagle Station (Kentucky) meteorite, contain iron-rich olivines and metals with compositions unlike those of other iron groups. It is therefore useful to distinguish between this "Eagle Station group" and "main-group" pallasites. These groups also differ in the oxygen isotopic compositions measured for their olivines. The two must have had origins in different parent bodies, although they look superficially similar.

MESOSIDERITES

Mesosiderites are stony-iron meteorites that have been aptly described as "wastebuckets" because they are agglomerations of such different materials. They consist of roughly equal proportions of metal and silicates. The chief problem in understanding these me-

Fig. 6.17. Mesosiderites are an important class of stony-iron meteorites. The Reckling Peak A79015 (Antarctica) mesosiderite contains dark gray silicates embedded in more highly reflective metal. The brecciated texture of this meteorite argues for its assembly after impact processes had mixed the two kinds of materials. The cube measures 1 centimeter on a side. Photograph courtesy of NASA.

teorites is that these two components apparently had nothing to do with each other – they are just accidental mixtures. Some of these mixed meteorites have discernible brecciated textures, as illustrated in Figure 6.17.

The silicate fraction of mesosiderites consists mostly of olivine, pyroxenes, and calcic plagioclase. These minerals and their compositions are very similar to those of the eucrite association, and the bulk chemical compositions of the silicate fractions are similar to those of howardites. Subtle chemical distinctions, however, preclude mesosiderites being simple mixtures of howardites plus metal. These chemical differences suggest that seasoning of the howardite-like mixture with an assortment of other rock types besides diogenites and eucrites was probably necessary.

Metals in mesosiderites have very uniform compositions, unlike the wide chemical ranges that have been seen in iron meteorites. This exotic metal must have had a history different from those of irons, but just what its origin was remains uncertain.

These two different kinds of materials are commonly shocked and were apparently physically mixed by impacts on their parent body's surface. The resulting breccias were apparently deeply buried and have been metamorphosed to varying degrees. Many mesosiderites are strongly recrystallized, now having only blurred images of their former textures. Slow cooling from peak metamorphic temperatures produced Widmanstatten patterns in metals. The cooling rates for mesosiderites measured from nickel profiles in taenite are exceptionally slow, less than half a degree per million years. No completely satisfactory explanation for how surface breccias could have been cooled so slowly has been proposed.

PRECIOUS METALS

The massive core of the earth is now and always will be totally beyond our scientific grasp, at least in terms of direct analysis of samples. However, among the more amazing gifts that nature bestows are occasional samples of the cores of other solar system bodies. Iron and stony-iron meteorites carry records of differentiation events probably much like those that produced the earth's interior.

The properties of iron meteorites suggest that they originally formed as pockets of liquid metal inside differentiated asteroids. The segregation of metal from silicates was largely complete in most cases, and the chemical trends observed in most irons match those predicted by laboratory crystallization experiments. A few iron groups appear to have been only partially melted, and thus these record the earliest stages of core formation. Cooling was slow enough to produce Widmanstatten patterns in irons that had the right chemical compositions. Oxygen isotopes in sparse silicate inclusions in a few iron groups suggest affinities with chondrite parent bodies. Differentiation probably happened early in the history of these bodies and may have accompanied the partial melting of silicate mantles to form achondrite magmas.

Studying the core-forming process as recorded in meteorites has been the focus of this chapter. In the following chapter we shall try to reconstruct some of the parent bodies for irons and stony-irons. Identification of possible examples of these bodies may also provide further insights into core formation.

SUGGESTED READINGS

The first reference provides an excellent introduction to iron meteorites and is profusely illustrated. The other works cited are demanding, but they provide an abundance of detailed information and interpretation of these objects.

GENERAL

Buchwald V. F. (1975) *Handbook of Iron Meteorites*, University of California Press, Berkeley, 1418 pp. (An exhaustive technical reference that provides a lucid overview of the field and individual descriptions for most iron meteorites.)

Scott E. R. D. (1979) Origin of iron meteorites. In *Asteroids*, edited by T. Gehrels, University of Arizona Press, Tucson, pp. 892–925. (Technical review that summarizes this complicated literature into a readily understandable form.)

CLASSIFICATION OF IRONS

Scott E. R. D. and Wasson J. T. (1975) Classification and properties of iron meteorites. *Reviews of Geophysics and Space Physics* 13, 527–546. (Technical review focusing on chemical classification of iron meteorites.)

Wasson J. T. (1974) *Meteorites*, Springer-Verlag, Berlin, 316 pp. (A technical monograph containing useful classification tables for most known iron meteorites.)

MINERALOGY OF IRONS

Buchwald V. F. (1977) The mineralogy of iron meteorites. *Philosophical Transactions of the Royal Society of London A* 286, 453–491. (Technical paper summarizing the many minerals that have been found in iron meteorites.)

STONY-IRONS

Buseck P. R. (1977) Pallasite meteorites – mineralogy, petrology and geochemistry. *Geochimica et Cosmochimica Acta* 41, 711–740. (Technical paper describing the properties of pallasites.)

Floran R. J. (1978) Silicate petrography, classification, and origin of the mesosiderites: Review and new observations. In *Proceedings of the 9th Lunar and Planetary Science Conference*, Pergamon Press, New York, pp. 1053–1081. (Technical paper focusing primarily on metamorphic effects in mesosiderites.)

THERMAL HISTORY

Wood J. A. (1964) The cooling rates and parent planets of several iron meteorites. *Icarus* 3, 429–459. (A classic technical paper that first outlined the principle of metallographic cooling rates.)

Narayan C. and Goldstein J. I. (1985) A major revision of iron meteorite cooling rates – an experimental study of the growth of the Widmanstatten pattern. *Geochimica et Cosmochimica Acta* 49, 397–410. (Technical paper that describes current thinking about cooling rates in iron meteorites.)

RELATIONS AMONG IRONS AND STONY-IRONS

Clayton R. N., Mayeda T. K:, Olsen E. J., and Prinz M. (1983) Oxygen isotope relationships in iron meteorites. *Earth and Planetary Science Letters* 65, 229–232. (Technical paper presenting oxygen isotopic data for silicate inclusions in iron meteorites.)

7 Iron and stony-iron parent bodies

The relative emptiness of space between 2.0 and 3.5 AU has been anything but a sanctuary for preservation of the smaller bodies of our solar system. The planetesimals that compose the present asteroid belt are mostly relics of some terrible destruction, a cosmic Armageddon that has left few bodies intact. Nowhere is this seen more clearly than in the parent bodies of irons and stony-irons. These meteorites are samples of the metallic skeletons that supported the rocky parts of differentiated asteroids. Their parent bodies must be cores from which the silicate mantles have now been stripped off and reduced to orbiting rubble.

In this chapter we shall attempt to predict the characteristics of such parent bodies. These properties will then be used to identify possible examples among the other debris littering the interplanetary battleground otherwise known as the asteroid belt.

CORES AND RAISINS

The cooling rates for iron and stony-iron meteorites require burial within objects no larger than respectable asteroids, as is illustrated in Figure 7.1. The vertical bars on the right side of this diagram show depth-of-burial conversions for planetesimals of various sizes. For example, the range of cooling rates for IAB irons could have been established at 30 kilometers depth in a 1,000-kilometer-diameter body or 50 kilometers depth in a 400-kilometer-diameter body, but even the very center of a 200-kilometer-diameter asteroid would have cooled faster. Cores buried to depths greater than 180 kilometers in a body 1,000 kilometers in diameter would not cool to 500°C (the temperature at which diffusion in metal effectively ceases) even within the age of the solar system, 4.6 billion years. The implication is that the parent bodies for all iron meteorites must have been relatively small, especially if irons formed within central cores. Some recent experiments on the effect of phosphorus on the growth of Widmanstatten patterns suggest that

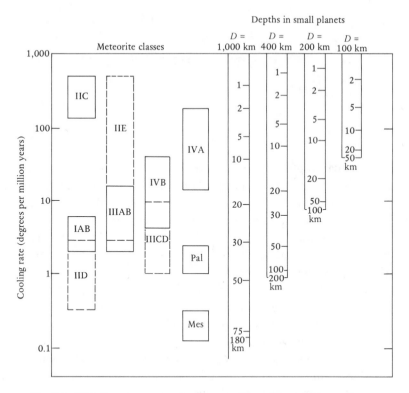

Fig. 7.1. This diagram summarizes the metallographic cooling rates for various groups of iron meteorites. Because the scale is logarithmic, groups that fall in the upper part of the figure (for example, IIC) show the most variation in cooling rate. The vertical scales on the right illustrate depth-of-burial conversions for asteroids of different sizes that would produce these cooling rates. Irons buried deeper than 180 kilometers in a 1,000– kilometer-diameter body would not cool even in 4.6 billion years. Therefore, the cooling rate data suggest that iron meteorites must have formed in the interiors of relatively small bodies or near the surfaces of somewhat larger bodies.

cooling rates should actually be several orders of magnitude faster. If these data are correct, either iron meteorite parent bodies need be only a few kilometers in diameter, or else iron meteorites could have formed near the surfaces of somewhat larger parent bodies.

Each of the individual iron groups represents a distinct chemical system, and therefore probably a different core. Within any central core it seems reasonable that fractional crystallization would result in some chemical variations, which in turn would produce mineralogical and structural differences, and indeed these are recorded. However, one observed variation among members of the same iron

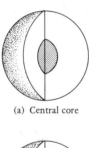

(a) Central core

(c) Fragmented and reaccreted core

(b) Failure of metallic pockets to congregate

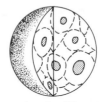

(d) Accretion of smaller differentiated bodies

Fig. 7.2. Possible models for iron meteorite parent bodies are summarized in these sketches: (a) a large central core, (b) dispersed masses of metal forming a raisin-bread structure, (c) a fragmented core that has been reaccreted into a second-generation asteroid, and (d) an agglomeration of smaller differentiated bodies. Cooling rate data for iron meteorites favor the last three models, unless parent bodies with central cores were very small.

group that is not expected is in cooling rate, because metal is such a good thermal conductor that the entire core should have cooled at approximately a uniform rate. Specimens of some iron groups show only a modest range of cooling rates, such as IIIAB (generally 1 to 10 degrees per million years), but others like IIC (15 to 250 degrees per million years) are highly variable. This has led to speculation that not all iron groups formed central cores. Instead, they may have been small nuggets of metal dispersed within parent bodies like raisins in a loaf of **raisin bread**. Each mass of metal would then have had its own cooling history, predicated on its burial depth. In this case, one iron group could contain samples from a number of these raisins, each with its own cooling history, but the parent body would have stamped its individual chemical fingerprint on all of these metal segregations.

Such raisin-bread parent bodies might have arisen in several ways. Localized melting could have produced pockets of liquid metal too small to congregate into a larger central core. An alternative is that differentiated parent bodies, complete with already-solidified but still hot cores, were catastrophically broken up and rapidly reaccreted before cooling of core fragments was complete. Yet a third

possibility is that core formation took place within small planetesimals that were subsequently accreted into larger composite bodies
before complete cooling.

It would, of course, be much easier for impacts to dislodge pieces
of metal if they were randomly embedded within a breakable silicate mantle than if they were part of a massive central core. Sampled parent bodies with raisin-bread structures could therefore still
be intact, although disruption of such bodies would probably be
necessary to obtain nuggets buried at many different depths. Sampling of central cores requires catastrophic destruction of at least
the rocky mantles surrounding them – there is no way around this
restriction. We are all familiar with the way the rounded yolks of
hardboiled eggs break cleanly away from their white envelopes. It
is tempting to think that something like this process has affected
many parent bodies with central cores. In such cases, these asteroidal relics would be bald metallic spheres, nearly devoid of ad-
hering silicates.

The evolution of mesosiderites involved mixing of unrelated silicates and metal to form regolith breccias, followed by deep burial,
as indicated by their extremely slow cooling rates (generally 0.1
degree per million years). This scenario is also consistent with the
assembly of small planetesimals into a larger body. In this particular example, the colliding planetesimals could have been of different types, one of metal and the other of unrelated silicate materials. Thus, there is evidence in mesosiderites to support the
allegation that their final residence was in a parent body with raisin bread structure, in this case assembled from the debris of previous impacts.

Raisin-bread models for core-containing asteroids may seem
rather contrived, but it should be noted that the same sort of history was intimated for many chondrite parent bodies. It has been
postulated that these were disrupted by impact and reaccreted to
form brecciated rubble piles. This kind of behavior may be the norm
in the asteroid belt.

THE CORE-MANTLE BOUNDARY

Pallasites are thought to have formed at the outer fringes of cores
where they were in contact with silicate mantle material, because
less dense olivine should have floated away from molten metallic
cores. The similarities in metal compositions and cooling rates of
main-group pallasites and IIIAB irons suggest that this particular
core was rimmed by these pallasites. The Eagle Station group of

pallasites is not obviously related to any known iron group, so it is conceivable that its parent body may not have been eroded down to the point where raw metal core was exposed. The close-packed arrays of olivine grains in pallasites indicate that they accumulated in this condition, and liquid metal filtered in to fill the spaces between them. For main-group pallasites this evidently happened after most of the metal core had already solidified, because the pallasite metal has a more highly fractionated composition relative to IIAB irons.

Let us examine a mechanism by which core and mantle materials could have mixed. The weight of a layer of olivine crystals accumulated from silicate magma at the core-mantle boundary would exert a downward force sufficient to submerge the lowermost olivines in the underlying molten metal, as illustrated by the left-hand column in Figure 7.3. Possibly a more realistic case, shown in this figure by the center column, occurs when the magma can erupt to the surface of the parent body. Under these conditions, unmelted crustal rock is not buoyantly supported by the magma beneath it, and its additional weight produces an even thicker layer of submerged olivines. In either case, pallasites would be formed at the interface between metal core and silicate mantle. The occurrence of at least two distinct groups of pallasites indicates that this boundary-layer process happened within more than one parent body and thus was not an isolated incident.

It appears that pallasites, like irons, could not have formed in bodies greater than several hundred kilometers in diameter. The increased pressures in larger bodies would have deformed olivines and squeezed out the interstitial molten metal, as illustrated in the right-hand column of Figure 7.3. However, the slow cooling rates measured for pallasites (0.5 to 2 degrees per million years) cannot be achieved in bodies any smaller than this. Figure 7.1 indicates that a body at least 400 or 500 kilometers in diameter would be necessary to produce these cooling rates. Either the calculated cooling rates are in error, because of some peculiarities in pallasite metal compositions, or these meteorites had to have formed in small planetesimals that subsequently joined to form larger bodies before complete cooling. The latter option would produce second-generation parent bodies with raisin-bread structure.

In order for core-mantle boundaries to have been sampled, pallasite parent bodies would have had to be substantially eroded. These would probably be similar to the metallic spheres discussed earlier, except that their surfaces would be studded with olivines. In the case of main-group pallasites, their parent body probably no

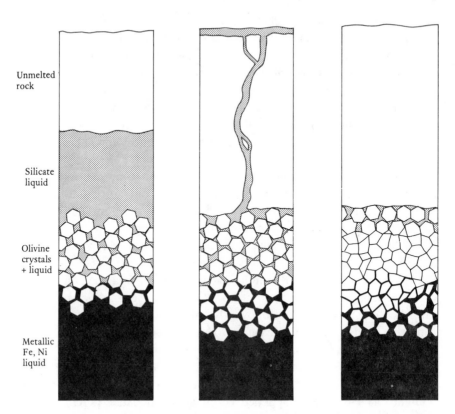

Unmelted
rock

Silicate
liquid

Olivine
crystals
+ liquid

Metallic
Fe, Ni
liquid

Fig. 7.3. These columns illustrate the possible formation of pallasites at a core-mantle boundary. In the left-hand column, the lowermost crystals in an accumulated pile of olivines are submerged in the still-molten core. In the center column, venting of magma to the surface allows the entire weight of overlying rock to press upon the olivines, causing even more crystals to be submerged. Subsequent solidification of metal in either case will produce pallasites at the interface. However, in large bodies, the increased pressure will cause olivines to deform, squeezing out interstitial liquid metal and preventing pallasite formation, as shown in the right-hand column.

longer exists, because the abundance and properties of the related IIIAB irons suggest that we have received a representative sample of its underlying core. The parent body for the Eagle Station pallasites could conceivably still be intact.

A CORNUCOPIA OF CORES

It is difficult to say how many parent bodies are represented by our present collections of iron and stony-iron meteorites, but the number must be fairly large. Each of the twelve groups of irons was certainly derived from a different core, presumably inside different parent bodies. Irons are classified as anomalous if they do not form a group of at least five members. However, some of these can be assigned to "grouplets" with several representatives, and the number of possible parent bodies expands to about sixty when these grouplets and unique irons are included. There are at least two pallasite parent bodies that have been sampled, although the main-group pallasites and group-IIIAB irons probably came from the same object. At least one mesosiderite planetesimal is also required.

This is a very generous number of planetesimals, especially when compared with the modest number of stony meteorite parent bodies that have apparently been sampled. Of course, it is possible that different chemical groups or grouplets formed within the same raisin-bread structured body, but this seems rather unlikely unless these bodies were second-generation objects that accreted from random, unrelated materials. In the latter case, we still require many small planetesimals to make cores in the first place. The explanation for the abundance of sampled iron meteorite parent bodies may be deceptively simple and apparently lies in the recognition that iron meteoroids are stronger than stony ones. As these objects orbit in space, they are subjected to a continual degradation process by mutual impacts. Irons are more likely to survive such encounters, and thus their parent bodies are better represented in meteorite collections.

Many of the iron parent bodies that have already been sampled almost certainly no longer exist. Sampling cores is not like sampling the near-surface lava flows of asteroid 4 Vesta. That we have these samples at all implies that their original parent bodies have been at least partly if not totally disrupted. Therefore, the chances of ever pinpointing the still orbiting body from which any specific iron group was derived seem remote.

SHINY BEADS

The spectrum of sunlight reflected from the surfaces of iron meteorites is frankly rather boring. Although these meteorites are highly reflective, they exhibit no peaks and valleys, the diagnostic absorption features on which asteroid spectral interpretation has come to depend. The gently sloping, almost linear spectra for four irons are illustrated in Figure 7.4. The only real variation in these is a de-

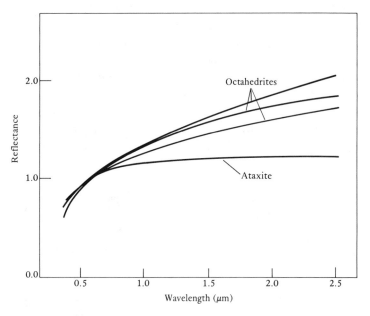

Fig. 7.4. *The reflectance spectra of three octahedrites and one ataxite are rather featureless. The ataxite has a flatter spectrum, which is characteristic of iron meteorites with higher nickel contents. These spectra are unfortunately not very useful for recognizing iron meteorite parent bodies by remote sensing.*

crease in slope with increasing nickel content. This is demonstrated by the flat spectrum of the ataxite, containing nearly 12 percent nickel, relative to those of the other three octahedrites in this figure.

However unexciting these spectra may be, the high albedos, coupled with the absence of absorption features attributable to silicates or opaque phases like carbon, may be at least partly diagnostic for metal-rich planetesimals. One class of asteroids, called the M type, exhibits this kind of sloping, featureless reflectance spectrum and may possibly be iron cores that have been stripped of their mantles. Incompletely denuded planetesimals or those with raisin-bread structures would contain differentiated silicates in addition to metal. These would probably have spectral signatures much like S-type asteroids. Earlier it was suggested that these might be ordinary chondrite parent bodies, but some combination of metal and silicate from differentiated asteroids would provide an equally acceptable model. Observed variations in the reflectance spectrum of one S-type asteroid, 8 Flora, as it rotated have been interpreted as evidence that such bodies are differentiated rather than being composed of homogeneous (by definition) chondrite. Even though M and S asteroids provide promising candidates for the parent bodies of irons, it is difficult to specify the proportion of metal in either type because of the featureless spectrum of this material. M types may have perhaps 10 to 100 percent metal on their surfaces, and S types could have 25 to 80 percent metal. Despite these uncomfortably wide ranges of possibilities, what is clear is that these spectral types of asteroids are the only reasonable candidates for iron meteorite parent bodies out there.

The reflectance spectra for stony-iron meteorites are very difficult to obtain in the laboratory, because it is virtually impossible to crush mixtures of metal and silicate uniformly. It may seem strange to worry about crushing malleable metal for this measurement; however, this material would be brittle at the frigid temperatures of interplanetary space, and its behavior when impacted on its parent body would probably differ significantly from that in a warm laboratory on earth. Because of this, any comparison of measured spectra for asteroids with stony-iron meteorites should be viewed with suspicion. A better approach is probably to calculate what the spectra would look like by numerically integrating the spectra of an octahedrite and either olivine (for pallasites) or a howardite (for mesosiderites). As might be predicted, such calculated spectra look much like those of irons, with small dips due to silicate absorption bands.

Some very respectable matches for these calculated reflectance patterns have now been discovered among asteroids. Potential parent bodies for pallasites (now called A type) have been identified among planetesimals previously classified as R type. Spectral matches for mesosiderites have also been located. The mixing of metal with silicates seems almost accidental in mesosiderites and could have occurred when an iron meteoroid impacted a Vesta-like surface and was mixed into the regolith. It has even been suggested that 4 Vesta itself could be the object from which mesosiderites were derived. Although the silicate fractions of mesosiderites are very similar to those of the eucrite association, there are several reasons to believe that 4 Vesta is not the mesosiderite parent body. The detailed spectral map of Vesta did not disclose any regions rich in metal, and the slow cooling rates of mesosiderites argue for deep burial, possibly in a second-generation body that has now been disrupted. Some of the smaller M-type objects may be remnants of such a body.

The identification of some possible asteroidal parent bodies for

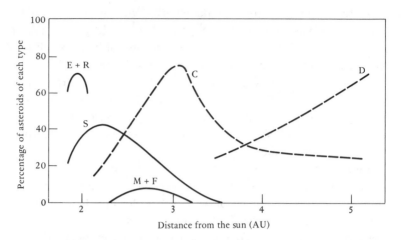

Fig. 7.5. The distributions for various spectral classes of asteroids suggest that the parent bodies for irons and stony-irons (possibly M, S, and some R types) are all located in the inner asteroid belt between about 2 and 3 AU. Did all of the bodies in this inner region partially melt and differentiate?

iron and stony-iron meteorites permits some inferences to be drawn about their occurrence in space. Figure 7.5 shows the distributions of various spectral types of asteroids in the asteroid belt. Part of this figure was presented earlier in our discussion of chondrite parent bodies, but M-type bodies were not included at that time. If irons and stony-irons are derived from M-, S-, and a subset of R-type asteroids, as has been argued by some experts in this field, then differentiated planetesimals completely dominate the inner asteroid belt. 4 Vesta also occurs in this region (at 2.36 AU) and may be the lone surviving relict from this original population of melted planetesimals. If this interpretation is correct, carbonaceous chondrites (C- and D-type asteroids) occur in the outer belt, and no obvious candidates for ordinary chondrite parent bodies exist anywhere in the belt. On the other hand, if some S-type asteroids are the parent bodies of ordinary chondrites, the distribution of differentiated and undifferentiated objects is more complicated. In this case, differentiated asteroids are still restricted to the inner belt, but not all inner-belt bodies have been melted.

M- and S-type asteroids also have different size distributions than C-type bodies. M asteroids are small, as is befitting stripped cores. The sizes of S-type planetesimals fall between those of C and M types, as is appropriate for only partly denuded cores.

ASTEROID FAMILIES

If we want to see the internal structure of an apple, we must first cut it open. Nature has already broken open asteroids for our perusal, but the job may have been done too well. The destruction that occurs when a planetesimal is subjected to impact produces large and small fragments derived from throughout the body. Reconstructing the original parent body from these is like putting together a puzzle face down, with no visual clues from the emerging picture on the front side. It is a difficult enough task just to recognize which fragments originally were parts of the same body. However, in a series of papers beginning in 1918, the Japanese astronomer K. Hirayama found a way to do just that. Hirayama surmised that the breakup of an asteroid into a collection of fragments, which he called a "family," would result in similar orbital characteristics for these bodies. In practice, corrections must also be made for minor orbital perturbations caused by nearby Jupiter. This results in what are called "proper" orbital characteristics. On the basis of similar proper orbits, Hirayama was able to recognize nine groups of asteroids, which we now call **Hirayama families**. He hypothesized that the members of any one family were collisional fragments of the same original planetesimal.

This idea has stood the test of time. One observation that supports this is that the orbital velocity differences among family members are small, generally 0.1 to 1.0 kilometer per second. These speeds are only a tiny fraction of their total orbital velocities and must approximate the original velocities at which they were ejected from their bursting parent bodies. More recently it has been determined that artificial satellites that have exploded in space have produced fragments with velocity distributions similar to those of Hirayama families. The effect of these small velocity differences over time is to spread out the family members along the original planetesimal's orbit, something like the dust trail left by a comet. In the years since Hirayama recognized his original nine families, many less conspicuous ones have been added, so that the count now stands at nearly a hundred.

Some indications of the compositions of family members can be obtained from their reflectance spectra. A convenient way of summarizing spectral observations is by plotting an average reflectivity (called the **color index**) for the ultraviolet-to-blue (short wavelength) end of the spectrum versus that for the blue-to-violet (long

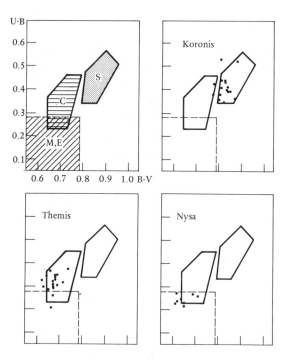

Fig. 7.6. *A plot of color index for ultraviolet-to-blue versus that for blue-to-violet offers a way to distinguish between asteroids of different spectral types. At the upper left are shown the fields for each type of asteroid. The other boxes contain data for asteroids in the Themis, Koronis, and Nysa families, groups of asteroids with similar orbital elements. The spectral similarities among the members of each family are evidence that each family represents fragments of a disrupted planetesimal. The Nysa family includes one large M object that may have been the core of its proto-asteroid.*

wavelength) range. Such a diagram, shown in Figure 7.6, resolves asteroids of different types, although there is some overlap between C, M, and E objects at the lower left corner. These three types can be readily distinguished if albedo data are available.

Also shown in the accompanying figures are spectra for asteroids in three different Hirayama families. The Themis family, named for asteroid 24 Themis, consists of C-type bodies. This family's home, like those of other C objects, is in the outer part of the asteroid belt. The original planetesimal has been estimated to have been about 300 kilometers in diameter, and the similarity of its members suggests that it had approximately the same composition throughout, as expected for a carbonaceous chondrite parent body. Members of the Koronis family, whose namesake is 158 Koronis,

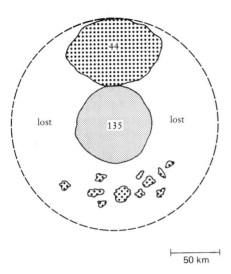

50 km

Fig. 7.7. A possible reconstruction of the original object from which the Nysa
family of asteroids were derived takes their spectral differences into account. 135
Hertha was possibly a metallic core surrounded by an achondrite mantle similar to 44 Nysa. Much of the missing rocky material has apparently been ground
into small chunks by impacts.

were originally part of a larger S-type asteroid. This body was also
homogeneous, indicating that it either was composed of ordinary
chondrite or was a corelike object whose mantle had even earlier
been removed. The Nysa family consists of two large asteroids about
70 kilometers in diameter – 44 Nysa, an E-type body, and 135
Hertha, an M type – and many small E objects about 20 kilometers
in diameter. A schematic family portrait before breakup is shown
in Figure 7.7. Hertha is assumed to have been a metallic core that
was surrounded by achondrite (in this case, aubrite, a class of
achondrites related to enstatite chondrites). The high ratio of core
material to mantle material that now characerizes the Nysa family
is consistent with collisional theory suggesting that stony materials
would have been broken into much smaller fragments than metal.

Not enough spectral data on families exist to make any definitive statements about the parent bodies of iron and stony-iron meteorites. However, the properties of the Nysa family support the
idea that cores in differentiated asteroids can be stripped of their
silicate mantles, and further links between these kinds of materials
should be sought. Future characterization of families by other observational techniques should provide additional important constraints on the internal workings of differentiation in small bodies.

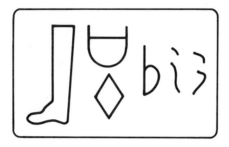

Fig. 7.8. These hieroglyphic symbols, found in Egyptian pyramids, mean "heavenly iron." They appear on the cover of the international journal of the Meteoritical Society. Courtesy of Meteoritics.

HEAVENLY IRONS

The Meteoritical Society, the international organization of meteo-riticists, imprints the hieroglyphic symbols shown in Figure 7.8 on the cover of its scientific journal. A literal translation of these ancient markings is "heavenly iron." Such symbols have been observed inside a number of Egyptian pyramids, where pharaohs were buried with artifacts constructed from iron meteorites. The desire to retain ownership of such objects in the afterlife did not stem solely from their utility or decorative appearance. From the name given to these metallic chunks it is clear that the ancient Egyptians recognized the uniqueness of their source and attributed some importance to it.

In this chapter we have attempted to fix the sources of iron and stony-iron meteorites much more precisely than did the Egyptians, but with only limited success. Iron meteorites clearly were formed in the cores of asteroids, but cooling rates for some groups suggest than their parent bodies may have contained dispersed minicores rather than massive concentrations of metal at their centers. Sampling of deep interior regions required the destruction of these bodies in most cases. Stony-irons, whether pallasites or mesosiderites, ultimately cooled in the deep interiors of asteroids that must have been similarly disrupted. Several spectral classes of asteroids may be possible candidates for denuded cores, but the featureless spectrum of iron-nickel metal makes quantitative estimation of metal contents impossible. These planetesimals are smaller than other asteroids, as we would expect for impact debris. The apparent restriction of these bodies to the inner asteroid belt suggests that thermal processing was important only in this region. The study of asteroidal families that represent disrupted parent bodies may offer some additional clues to understanding the insides of differentiated planetesimals.

The sources for iron and stony-iron meteorites still are not well constrained, as is indicated by the brevity of this chapter. Their importance as the only available samples that document core-forming processes nevertheless remains unchallenged. Heavenly irons provide real insights into the deepest reaches of our own and other planets.

SUGGESTED READINGS

The topic of iron meteorite parent bodies has been addressed infrequently in technical scientific literature, and never (to my knowledge) in a popular paper or book. The references below make challenging reading, but are the only sources available.

GENERAL

Scott E. R. D. (1977) Origin of iron meteorites. In *Asteroids*, edited by T. Gehrels, University of Arizona Press, Tucson, pp. 892–925. (Technical paper that describes the general properties of parent bodies as inferred from the properties of iron meteorites.)

REFLECTANCE SPECTRA AND DISTRIBUTION

Chapman C. R. (1976) Asteroids as meteorite parent bodies: The astronomical perspective. *Geochimica et Cosmochimica Acta* 40, 701–719. (Technical paper outlining the spectral characteristics of asteroids.)

Gradie J. C. and Tadesko E. (1982) Compositional structure of the asteroid belt. *Science* 216, 1405–1407 (Technical paper in which the locations of various asteroid types are summarized.)

Gradie J. C., Chapman C. R., and Williams J. G. (1979) Families of minor planets. In *Asteroids*, edited by T. Gehrels, University of Arizona Press, Tucson, pp. 359–390. (Technical paper discussing the properties of recognized Hirayama families.)

PARENT BODIES FOR IRONS

Goldstein J. I. and Short J. M. (1967) The iron meteorites, their thermal history and parent bodies. *Geochimica et Cosmochimica Acta* 31, 1733–1770. (Technical paper that uses metallographic cooling rates to model the sizes of iron meteorite parent bodies.)

Wood J. A. (1979) Review of the metallographic cooling rates of meteorites and a new model for the planetesimals in which they formed. In *Asteroids*, edited by T. Gehrels, University of Arizon Press, Tucson, pp. 849–891. (Technical review summarizing and interpreting available cooling rate data for iron meteorites.)

PARENT BODIES FOR STONY IRONS

Hewins R. H. (1983) Impact versus internal origins for mesosiderites. *Proceedings of the 14th Lunar and Planetary Science Conference, Journal of Geophysical Research, Supplement*, B257–266. (Technical paper discussing the properties of possible mesosiderite parent bodies.)

Wood J. A. (1978) Nature and evolution of the meteorite parent bodies: Evidence from petrology and mineralogy. In *Asteroids: An Exploration Assessment*, edited by D. Morrison and W. C. Wells, NASA conference publication 2053, pp. 45–55. (Technical paper in which the formation of pallasites at core-mantle interfaces is explored.)

8 A space odyssey

There is an anecdote, well worn with many variations, in which a grizzled denizen of rural Maine or some such place is asked to provide directions to a local landmark. After a number of abortive attempts, he is finally forced to the conclusion that "you can't get there from here." The very fact that we have meteorites on earth shows that this punch line does not apply to their parent bodies. The extraction of meteoroids from asteroids or planets and their delivery to earth is no laughing matter, and the difficulties and complexities involved in this feat would wither the resolve of anyone attempting to give explicit directions.

In the previous chapters we have dissected various kinds of meteorites and attempted to explore their parent bodies. The intent of this chapter is to tie up one monumental loose end − how these meteorites got from their parent bodies to the earth. The delivery of meteorites is, of course, only a minor by-product of the clockwork of the solar system. These celestial workings are vastly more intricate than the working parts of timepieces and cannot be observed directly; however, the mechanisms by which samples are liberated from their parent objects and the routes by which they find their way to earth can be analyzed indirectly. It is not possible to specify precisely the odyssey of any particular meteorite, but as we shall see, plausible paths can be construed.

ASTEROIDAL TRAFFIC ACCIDENTS

Typical relative velocities for asteroids encountering other asteroids are on the order of 5 kilometers per second. We cannot recreate in laboratory experiments what happens when two planetesimals collide at these speeds. The disastrous outcome can be determined only by extrapolating the results of fragmentation experiments at much lower velocities, but studies of nuclear weapons tests provide some basis for believing that these calculations are correct.

It probably comes as no surprise to the reader that such accidental collisions liberate meteoroids from their parent bodies. Most of the energy of the original projectiles gets transformed into heat or does the work of asteroidal disruption; however, as much as 10 percent of the original energy is transmuted into energy of motion for the resulting debris in cratering events. This kinetic energy provides a way for fragments to be cast free of their asteroidal moorings. If the energy of motion for a fragment exceeds the gravitational pull that binds the planetesimal together, the chip is said to have achieved **escape velocity**, and an independently orbiting meteoroid is born.

The brecciated nature of many meteorites demonstrates the role that impact processes have played in asteroidal evolution. We have even surmised that the internal structures of many asteroidal parent bodies are possibly like reaccreted piles of rubble, based on inconsistencies between their cooling rates and grades of metamorphism. Images of the cratered surfaces of the tiny Martian moons Phobos and Deimos, probably captured asteroids, also show the scars of numerous impact events in their past. Virtually every planetesimal in the asteroid belt has probably suffered impacts on some scale over the 4.6 billion years of its existence, and much of the present belt may now simply be battered collisional remnants. Thus, there have been ample opportunities for meteoroids to have been extracted from asteroids by mutual collisions.

THE PROPERTIES OF ORBITS

Ejection of pieces of rock or metal from their asteroidal parent bodies is the crucial first step in the long trek toward earth. However, completion of this trip requires that the meteoroids be placed into orbits that intersect the orbital path of their eventual planetary target. In order to understand how this hurdle is surmounted, we must first examine the properties of meteoroid orbits.

The trajectory followed by the original asteroidal parent body traces out an ellipse, although it may look superficially like a circle. A similar elliptical path is followed by any small object revolving about a more massive body, whether it is a large planet orbiting about the sun or a tiny spacecraft orbiting about the earth. The three most important "orbital elements" that define the size, shape, and orientation of the ellipse are its **semi-major axis** (the greatest distance from the center of the ellipse to its periphery), its **eccentricity** (a measure of its departure from circularity), and its **incli-**

nation (the angle between the orbital plane and that of the earth). The orbital **period**, the time required for completion of one orbital loop, is another useful parameter.

Although the orbital behavior of a small body may be described approximately by the geometry of an ellipse, the situation is actually somewhat more complicated. Often the gravitational effects of a third body also come into play. For example, the orbit of a spacecraft revolving around the earth may be modified at some points by the gravitational pull of the moon. These **perturbations** result in minor sinuosities being superimposed on an otherwise perfect elliptical track. Unfortunately, whereas we have equations that specify the geometry of an elliptical orbit for the two-body problem, there is no general mathematical solution that describes the motions when three bodies are involved. However, these perturbations can be predicted by computer simulation. The calculation procedure is tedious, involving many small steps. The forces are determined at each step, and the small body of interest is repositioned (perturbed) according to the results of the previous step.

In order for a fragment of an asteroid to become a meteorite, its orbit must be perturbed in such a way that it crosses the orbital path of the earth. This requires a significant increase in eccentricity. This is all well and good, except that the impact events that freed meteoroid fragments from their asteroidal parent bodies in most cases cannot have altered their orbits this drastically. Recall that in a previous discussion of asteroid families, we learned that most of the debris from a collision spreads out along the orbital path of the original parent object. Collisional shocks capable of producing more than small, say 0.1 AU, changes in semi-major axis would probably also grind the liberated fragments to dust. The highly elliptical meteoroid orbits required for earth capture have to be produced by some other kind of perturbing force. The next leg of the journey depends on this mysterious process. But what kind of force could do this?

GEOGRAPHY OF THE ASTEROID BELT

The position of planetesimals within the asteroid belt offers an important clue to the identity of this perturbing force. The distribution pattern of planetesimals is anything but smooth. This is clearly shown in Figure 8.1, a plot of the semi-major axes of the orbits of the first 400 numbered asteroids. This snaggletoothed distribution pattern was first noticed by the astronomer D. Kirkwood in 1867.

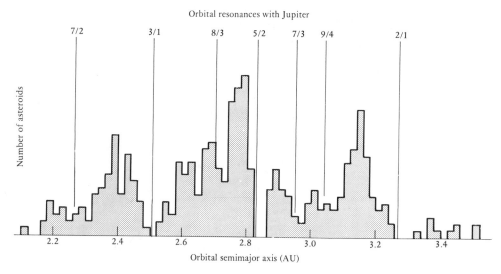

Fig. 8.1. The irregular distribution with distance from the sun of the first 400 numbered asteroids is clear from this diagram. Vacancies, called Kirkwood gaps, in the asteroid belt are due to commensurabilities with the orbit of Jupiter, the locations of which are marked by vertical lines. The fraction 7/2 means that an asteroid in that position makes 7 revolutions around the sun for each 2 revolutions completed by Jupiter. Asteroids at such resonant positions are affected by Jupiter's massive gravitational field more often than those in other, random positions. In this way, objects temporarily stored in these gaps may be perturbed into earth-crossing orbits.

The relatively empty spaces in the asteroid belt are now called **Kirkwood gaps**. These are not actually void zones, because asteroids with elliptical orbits pass through them; however, the gaps are regions in which almost no asteroids remain for long.

Although the Kirkwood gaps seem to occur at random distances from the sun in this figure, they are in fact at very specific locations relative to the orbit of Jupiter. Each of these localities corresponds to a **commensurability**. An asteroid's orbit is commensurable if it has some periodicity with that of Jupiter, that is, when the orbital periods have a ratio of small whole numbers. One example would be an asteroid whose period is one-half that of Jupiter. On every second revolution around the sun, this asteroid will find itself adjacent to the giant planet. In Figure 8.1, commensurabilities are marked by appropriate fractions. The numerator in each fraction refers to the number of revolutions an asteroid at that position must make in order to line up with Jupiter after it has made the number of revolutions given by the denominator. The blatant cor-

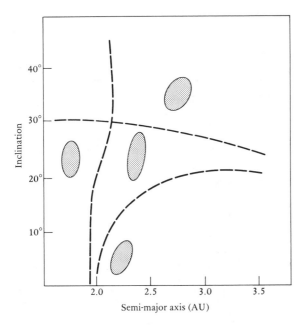

Fig. 8.2. In addition to having gaps in semi-major axis, asteroids also avoid certain inclination positions. The dashed lines in this figure are ''surfaces'' of resonance, actually combinations of inclination and semi-major axis along which asteroids are almost absent. The locations of concentrations of asteroids are shown by ovals. Meteoroids thrown onto one of these resonant surfaces can be perturbed into earth-crossing orbits.

respondence between commensurabilities and the positions of Kirkwood gaps indicates some causative link.

Commensurabilities are also associated with gaps in inclination, in addition to those in semi-major axis just described. In Figure 8.2, the curved lines represent combinations of inclination values and semi-major axis that are commensurable with Jupiter's orbit.

What produces these gaps in distribution? Because asteroids in commensurable positions experience close encounters with Jupiter more often than do asteroids in other, random locations, they also feel the powerful tug of Jupiter's gravity more frequently. The repetition of the resulting perturbations creates a condition known as **resonance**. Any child learns that a swing must be pushed repeatedly in a manner that matches the swing's natural frequency. When this is done, the pushes are additive, and the swing's oscillations increase or at least stay constant. If the swing is pushed unevenly, its motion is damped. Resonances can act in a similar, though more spectacular, manner to change orbital characteristics.

This interaction between asteroids and Jupiter must somehow account for the formation of these gaps. The frequent gravitational tugs either swept commensurate areas clean or prevented the accretion of asteroids there in the first place. In any case, the behavior of any new meteoroids that wander into such inhospitable locations can be calculated. Gravitational interaction with Jupiter over time forces a marked increase in orbital eccentricity, resulting either in ejection from the solar system or in the less drastic effect of producing trajectories that pass into the inner solar system. Impact fragments spalled off asteroids and thrown into nearby Kirkwood gaps or inclination resonances may thus evolve into meteoroids with highly eccentric orbits. The commensurabilities are apparently open windows through which fragmental materials escape the asteroid belt. Orbiting in the gravitational shadow of giant Jupiter is obviously a precarious pastime.

THE PLANETARY PRISON

The kinetic-energy requirements to liberate small fragments from planetesimals are really quite modest. In fact, some future astronaut standing on the surface of a small asteroid could possibly launch a rock to escape velocity with a baseball throw. Although moons and planets experience the same buffeting by impacts that asteroids experience, their massive gravity fields exert a near stranglehold on the larger fragments produced during cratering. It has generally been supposed that any smaller fragments that could be ejected from planets by impact mechanisms would have experienced such a high degree of shock that they would be pulverized, melted, or even vaporized. Yet no other natural means of meteoroid ejection seems possible. The energy of rapidly expanding gases during volcanic eruptions is too small to accelerate fragments to planetary escape velocities, and other geologic phenomena are even less capable launching mechanisms. How, then, have we received intact pieces of the moon and possibly Mars?

The answer, at least in part, may be provided by calculations in which the differences in impact behaviors of target materials near the ground surface versus those at depth are considered. The disturbance created by an impacting object propagates through the subsurface as a stress wave whose force lessens as it moves away from the impact site, like the expanding wave produced when a pebble is thrown into a pond. Target rocks close to the impact site are melted or pulverized into dust or small grains, the fragment

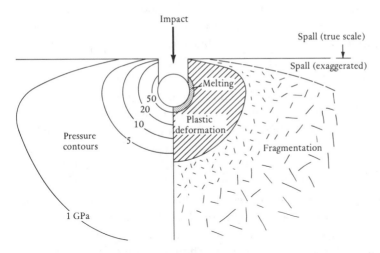

Fig. 8.3. This sketch schematically illustrates the effects of a major impact onto a planetary surface. The impacting object, of course, will be pulverized or vaporized. The left half of the figure shows contours (in gigapascals) of the shock pressures experienced at various distances from the impact site. The right side illustrates the effects produced in the target rocks. Close to the impact site, melting and/or pulverization of rocks take place, but farther away the rocks are broken into larger fragments. At the ground surface, virtually unshocked fragments spall off and can be accelerated to planetary escape velocities.

size increasing away from ground zero, as illustrated in Figure 8.3. However, rocks very near the ground surface experience several kinds of shock waves that partially cancel each other. This area of wave interference offers a shelter from the full force of the shock wave. Calculations indicate that some of this near-surface material will spall off as relatively unshocked fragments and can be accelerated to high speeds. Limited shock is important in the case of the lunar meteorites, which have experienced only minor shock metamorphism.

The only potential problem with this explanation is that the chips ejected at planetary escape velocities must be rather small. The lunar meteorites are approximately the size of golf balls, and calculations indicate that these could readily achieve lunar escape velocity (2.4 kilometers per second). But most of the shergottites are larger, up to grapefruit sizes, and these may have lost half their mass during atmospheric transit. The largest crater in the "young" terrane of Mars, a necessary location from which to derive shergottites with young crystallization ages, is 30 kilometers in diameter. The maximum size for a fragment that could be accelerated to Martian escape velocity (about 5 kilometers per second) from such a crater is approximately 1 meter. Yet the various shergottites are different enough that they almost certainly did not all occur in a 1-cubic-meter volume. However, it is not hard to conceive that a number of sub-meter-sized chunks from the periphery of such a crater could have been ejected during the same event.

It is also conceivable that Martian meteoroids have been ejected by some other mechanism. One alternative idea was prompted by indications of a subsurface permafrost (ice) layer on Mars. It has been hypothesized that these ices would vaporize on impact, providing a kind of jet-propulsion mechanism for fragments of overlying rock. A second suggestion utilized oblique impacts into the Martian surface. Ejecta fragments 10 to 100 meters in size could possibly be entrained behind a ricocheting projectile and accelerated to Martian escape velocity. There is some observational evidence that a few large craters on the Martian surface were produced by oblique impacts.

The mechanisms by which rocks escape from planetary gravity fields are not well understood. However, it now seems likely that this does happen, and the ideas presented earlier are reasonable attempts to understand the process. Once free of the gravitational grasp of a moon or planet, a meteoroid continues to travel alongside the larger body. The orbit of a Mars fragment would be so perturbed by gravitational interaction with that planet that it could readily become more eccentric and approach the earth. An ejected lunar sample has an even higher probability of being swept up by the earth because of its proximity. Thus, these meteoroids do not require the windows opened by the gravitational field of Jupiter in order to make good their escape.

ATEN, APOLLO, AND AMOR

A number of presently observable asteroids have orbits that periodically cross the earth's orbital path. These small bodies are typically only a few kilometers or less in diameter. Objects like these or fragments chipped from them are obvious candidates for meteorites, because they are already traveling on potential collision courses.

The earth's elliptical racetrack brings it as close to the sun as 0.983 AU and takes it as far out as 1.017 AU. Earth-approaching asteroids are classified by their orbital dimensions relative to that of the earth. **Aten asteroids** are those whose orbits remain inside

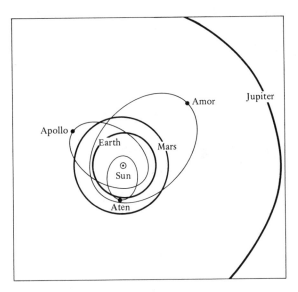

Fig. 8.4. The ellipses illustrate orbits for three kinds of earth-approaching asteroids. Aten asteroids have semi-major axes less than 1.0 AU, whereas those for Apollo asteroids are greater than 1.0 AU. Both follow orbital paths that intersect the earth's orbit. Amor asteroids make only close approaches to the earth's orbit, but periodic perturbations cause them to evolve into Apollo objects.

the earth's path most of the time. Their semi-major axes are less than 1.0 AU, so that they have the potential to impact the earth only when they are at or near their orbital extremities. In contrast, **Apollo asteroids** spend most of their time outside of the earth's orbital path, but at some point of close approach to the sun, they cross the orbit of the earth. **Amor asteroids** make relatively close (within about 0.3 AU) approaches to the earth's orbit, but do not actually overlap it. Schematic orbits for these three classes of asteroids are illustrated in Figure 8.4. The distinction between Apollo and Amor objects is really rather arbitrary, because these kinds of asteroids can and do switch back and forth by means of periodic perturbations. As an example, 1915 Quetzalcoatl is classed as an Amor asteroid because it does not presently cross the earth's path; however, analysis of its past orbital evolution indicates that it was an Apollo object prior to the year 1943. The Amor asteroids are therefore lumped for consideration with the earth-crossing objects.

There are approximately 50 earth-approaching objects whose orbits have been fairly well documented. The total number of these asteroids, most of which have not yet been discovered because of their tiny sizes, has been estimated to be on the order of 1,300, of which a small fraction (approximately 8 percent) are Atens, roughly half are Apollos, and the remainder are Amors. This is a sizable number of potential little earth smashers. It is clear that these earth-crossing asteroids cannot have been permanent residents of the inner solar system. From the number of Apollo asteroids and an estimated collision rate, it is possible to extrapolate back in time, and it is found that trillions of former asteroids with earth-crossing orbits would be required to account for the present population. The total mass of these hypothetical objects would have been 1,000 times that of the sun! Moreover, the initial population of these objects would have had to diminish steadily throughout geologic time. The impact records of the earth and moon indicate no such steady decline during the last 3 billion years.

It is not possible to calculate the probability that any one of these objects will impact the earth, because close encounters with planets constantly scramble their orbits. These nomads suffer many such orbital modifications before they meet their ultimate destiny – collision with a planet or ejection from the solar system. It is possible, however, to calculate the average lifetimes for earth-crossing asteroids as a group. Computer programs can be devised to simulate random encounters with planets and to estimate how long asteroids could remain in such orbits before oblivion. This analysis technique is really a form of statistical roulette and has appropriately been dubbed the "Monte Carlo" method. Mean orbital lifetimes determined for these objects are on the order of only 10 to 100 million years. Therefore, earth-crossing asteroids must have been injected into their present orbits in relatively recent times.

One way to do this, of course, is by employing the perturbations inflicted by Jupiter at resonance locations in the asteroid belt. However, estimates of the yield of objects from these gaps are apparently too low to account for all of the earth-crossers. For this reason, it seems probable that at least some of these bodies are the nuclei of extinct comets. After a number of close passes near the sun, short-period comets will lose their ices, and possibly all that will remain will be rocky residues. Perturbations by planets alter cometary orbits in such a way that they become earth-crossing.

Reflection spectra measured for earth-crossing asteroids indicate that many of the diverse compositional types found in the main asteroid belt are present in the earth-crossing population as well. For example, 1915 Quetzalcoatl is spectrally similar to 4 Vesta and eucrites; 1862 Apollo is classified as an S asteroid, possibly an ordinary chondrite or stony-iron; and 2100 Ra-Shalom is a C type that compares favorably with carbonaceous chondrites. All of these could be chunks of still existing or now-disrupted planetesimals in the main belt. 2201 Oljato may be an example of the nucleus of a near-dead comet. It is not a luminous body, but particles are apparently being ejected from its surface, a behavioral pattern characteristic of active comets.

Aten, Apollo, and Amor objects almost certainly represent the last stage of a meteoroid's journey to earth. At least some of these have probably already been sampled, and others are just meteorites waiting to happen.

METEOROIDS EXPOSED

When stimulated by light, a chemical reaction on plastic film produces a photographic negative. In a somewhat analogous manner, meteoroids orbiting in space act as film when exposed, in this case to cosmic rays. The reactions resulting from this irradiation provide an indelible record of their travels.

When cosmic rays strike the nuclei of atoms in orbiting pieces of rock or metal, they tend to dislodge protons and neutrons, a process known as **spallation**. The reactions change the identities of the target atoms; for example, the removal of one proton and two neutrons from ^{56}Fe (an isotope of iron) produces ^{53}Mn (an isotope of manganese). In some cases the newly created isotopes are sufficiently different from what already composes the object that they can be recognized and analyzed. Spallation-produced isotopes can be either stable or radioactive, but the radioactive ones are much easier to measure.

Cosmic rays are particles with a variety of energies. Those from solar wind and flares are fairly weak and cannot effectively bore into rock or metal, but more energetic galactic cosmic rays can penetrate to depths of about a meter. Thus, only the outermost meter of large meteoroids, really negligible amounts of material, will experience spallation reactions. Put another way, most orbiting meteoroids will not be exposed to cosmic radiation until they are broken into fragments of meter size or smaller. Whenever this happens, they begin to accumulate spallation products. The time interval calculated from measurements of the accumulated radionuclides produced by cosmic-ray exposure, called the exposure age, then represents how long a meteorite orbited while within a me-

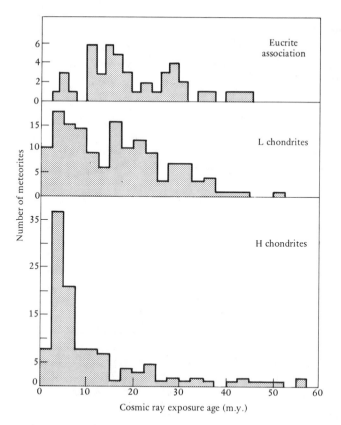

Fig. 8.5. Cosmic-ray exposure ages for various stony meteorite classes reveal different patterns. The large peak for H chondrites suggests that these meteorites were produced by a single collision that fragmented a large mass about 5 million years ago. Other classes of chondrites and samples from the eucrite parent body do not show such a pronounced peak in their exposure ages and may have been chipped off larger objects by numerous small impacts during the last 50 million years or so.

ter-sized or smaller object. Exposure ages are in most cases only approximations, probably accurate within a factor of 2, because so many assumptions are involved in calculating them. Nevertheless, they provide otherwise unobtainable information on the lifetimes of small meteoroids orbiting in space.

Cosmic-ray exposure ages have now been measured for a large number of meteorites. These measured values agree more or less with the orbital lifetimes for earth-crossing objects determined from Monte Carlo calculations (generally less than 100 million years). One of the interesting observations from these data is that expo-

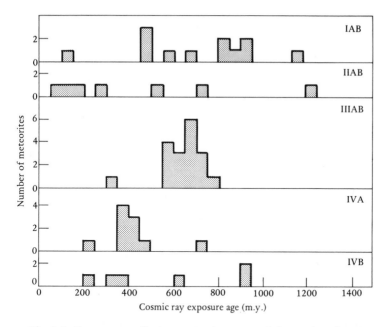

Fig. 8.6. Exposure ages for iron meteorites are much longer than for stony meteorites. Compare the horizontal scale for this figure with that of Figure 8.5 – all of the data in the previous figure would plot on the extreme left margin of this one. This difference presumably reflects the ability of small iron meteoroids to survive impacts in space.

sure ages for some meteorite groups tend to cluster about preferred values. For example, H chondrites have a distinct exposure-age peak at about 5 million years, as shown in Figure 8.5. Such clustering suggests that a single collision broke up a large mass to produce all the meteorites in the group. In contrast to the situation for H chondrites, exposure ages for L chondrites and members of the eucrite association are spread over about a 50 million year range, as shown in Figure 8.5. Meter-sized chunks of these meteorites were probably produced by successions of smaller impacts at various times.

Figure 8.6 demonstrates that iron meteorites have much longer exposure ages than their stony counterparts. (Note the change in scale – every one of the ages in the previous figure would plot at the extreme left margin of this figure.) Why would irons exist as small pieces before arrival on earth so much longer than stony meteorites? The difference in exposure ages probably simply reflects the greater strength of iron meteoroids relative to rocky bodies. Irons could presumably survive small impacts that stones could not, preventing them from being as readily destroyed in space. This difference in destructibility may explain why meter-sized fragments of irons can apparently persist in space for over a billion years, but only the most recently broken small pieces of stones can complete the trip to earth. The inner solar system can be thought of as a kind of Khyber Pass for meteoroids, with other orbiting masses waiting in ambush to destroy them.

Two groups of iron meteorites, IIIAB and IVA, show exposure age peaks and were probably produced by catastrophic disruptions. Other groups such as IAB or IIAB may have been broken off larger bodies at various times over the last billion years. This may mean that iron groups with clustered exposure ages formed in central cores, and those with large age spreads represent a series of raisins that were excavated at various times.

AT THE FINISH LINE

A runner's heartbeat and respiration rate continue at accelerated levels even after the race is finished, but will gradually slow to normal levels after a few minutes at rest. Meteorites likewise take time to recover from their interplanetary exertions, at least in terms of the decay times of the new radionuclides produced by cosmic irradiation. This can be a useful tool to discover how long meteorite finds have been on earth. The time of earth residence is called the **terrestrial age** of a meteorite.

After a meteorite falls, it is shielded from further exposure to cosmic rays by the earth's atmosphere. Consequently, no new isotopes will be made by spallation, only by the decay of the radionuclides already present. If the amount of a certain isotope produced by cosmic irradiation is measured in a recently fallen meteorite, we would expect that other spallation-produced nuclides should be present in amounts appropriate to the same duration of exposure. However, in many finds, the concentrations of short-lived radionuclides are found to be significantly less than predicted. The reason for this discrepancy is that the short-lived isotopes have decayed during the meteorite's residence on earth, but longer-lived radioactive isotopes produced by spallation are still present at or near their original abundances. Figure 8.7 illustrates this effect for chondrites found in Antarctica versus chondrite finds from other locations. The amounts of ^{53}Mn, an isotope of manganese produced by spallation reactions in space, are ap-

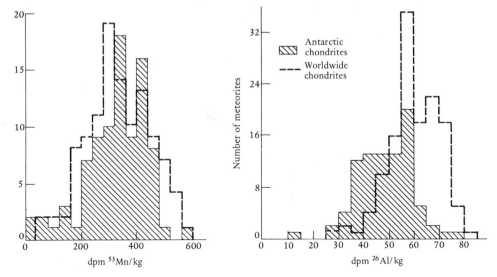

Fig. 8.7. The terrestrial ages of meteorite falls, the length of time they have been on earth, can be estimated from the relative decays of several radionuclides with different half-lives that were produced by cosmic-ray exposure. These two diagrams compare the radioactivities produced by ^{53}Mn and ^{26}Al (both expressed in disintegrations per minute per kilogram of sample). Data for chondrites found in Antarctica and elsewhere are presented. The amounts of ^{53}Mn are roughly the same in both Antarctic and non-Antarctic finds, but the amount of ^{26}Al is lower in Antarctic meteorites. ^{26}Al decays more rapidly than ^{53}Mn, so this difference reflects the longer terrestrial residence of meteorites in the Antarctic environment.

proximately the same in both populations. The amounts of ^{26}Al, an isotope of aluminum that decays more rapidly that ^{53}Mn, are generally lower in Antarctic chondrites.

From this simple comparison we can deduce that Antarctic meteorites, as a group, must have been frozen in the ice sheet for longer times than the terrestrial residence of non-Antarctic finds. This is one of the important features of Antarctic meteorites – they represent a population of objects that orbited in space at earlier times, prehistoric in human terms. Using several different spallogenic isotopes that decay at different rates, it is possible to quantify terrestrial ages for meteorites. Calculated terrestrial ages for finds from different regions of Antarctica appear to be somewhat different: about 100,000 years for Yamato Mountains specimens from Queen Maud Land, and about 300,000 years for Allan Hills specimens from Victoria Land. This variation is probably due to ice movements and different ablation rates, because all the meteorites at one locality were certainly not parts of a single gigantic fall.

AN OVERVIEW OF METEORITE HISTORY

The amount of progress made in understanding meteorites since the birth of this science is evident in the preceding chapters. This section is an attempt to synthesize meteorite evolution. We must also employ all of the chronological information at our disposal to unravel this history. Isotopic dating techniques record many different kinds of events, and it may be useful to summarize these. Decay of long-lived radionuclides generally is used to date the time of formation of parent bodies. In some cases, these isotope clocks have been reset by later metamorphism or igneous events, but for asteroidal bodies these events followed so closely on the heels of accretion that the differences are inconsequential. Other radiometric clocks were set in motion by later regolith-forming episodes on parent bodies. One example is provided by the isotopes in glasses produced from impact melts, which provide dates for such impact gardening. For instance, the ages for solidification of glass fragments in the Malvern (South Africa) howardite indicate that regolith formation continued on the eucrite parent body until at least 3.6 billion years ago. The youngest regolith event found so far in any meteorite is 1.4 billion years. If the daughter isotope in a dating scheme is a gas, it may be lost during major impact events, completely resetting that particular clock. Such degassing ages apparently record the times of ejection from parent bodies. As an example, the loss of an isotope of argon gas (^{40}Ar, a decay product of potassium) has been used to determine the times at which chondrites were thrown free of asteroids. Cosmic-ray exposure ages record the times at which orbiting meteoroids were subsequently broken into meter-sized fragments.

The events leading ultimately to the delivery of pieces of asteroids to the earth as meteorites are summarized in Figure 8.8. Parent-body accretion occurred approximately 4.6 billion years ago. Primitive (chondritic) planetesimals experienced only mild heating, resulting in metamorphism between 4.4 and 4.6 billion years ago. Processed (achondritic) asteroids were heated to the point of partial melting and experienced differentiation to form metallic cores at about the same time. Between 4.4 and 1.4(?) billion years ago, brecciation occurred on both primitive and processed asteroids. Impacts of small projectiles produced finely comminuted regoliths.

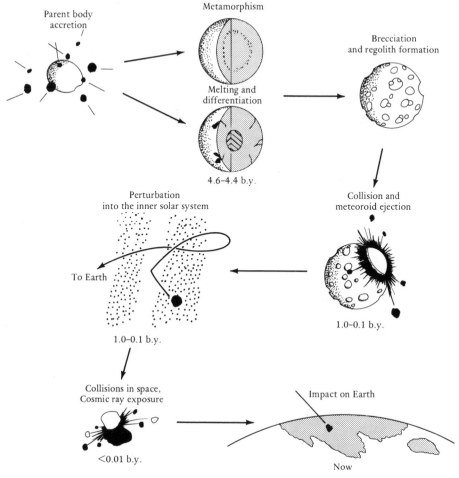

Fig. 8.8. This diagram summarizes the evolution of asteroidal bodies and the events that led to the delivery of samples of them to earth.

Collisions between bodies of comparable sizes may have caused the disruption of planetesimals, with subsequent reaccretion to form rubble piles. Measured collisional shock ages for meteorites range from 1.0 to 0.1 billion years. Meteoroids liberated from their parent bodies during such events may have been injected into resonant orbits with Jupiter and were subsequently perturbed into the inner solar system. Small pieces of iron cores ejected about a billion years ago could have survived intact to fall to earth, because cosmic-ray exposure ages for a few irons are this long; however, many irons and all stony meteorites must have been broken into smaller pieces by impacts in space within the last few tens of millions of years to account for their shorter exposure ages.

The migratory history of fragments from planetary bodies might be illustrated by the shergottites. We shall make the assumption that these are Martian samples, though that is irrelevant to the chronology of events outlined in Figure 8.9. Igneous activity continued at least until 1.3 billion years ago and possibly more recently, as indicated by the crystallization ages of shergottites and their relatives. The parent body suffered a major cratering event 180 million years ago, resulting in the resetting of some isotopic clocks by shock. It is not clear whether or not this event ejected shergottite fragments from their parent body. If it did, the fragments had to have been fairly large, because cosmic-ray exposure times for these meteorites are much shorter. If the cosmic-ray exposures ages (0.6 to 11 million years) date the times of ejection, the fragments could have been small, as seems likely from cratering theory. Calculations indicate that some fraction of Mars ejecta would be perturbed into earth-crossing orbits by close Martian encounters and reach its eventual target within the 11 million years required by the cosmic-ray exposure ages.

These historical sketches are little more than cartoons, but they represent a prodigious amount of research effort expended to reach this level of understanding.

ON THE IMPORTANCE OF METEORITES

The rich sound of a pealing church bell is due to its many vibrational modes, each ringing out at a different harmonic frequency. Similar modes occur in any vibrating metal plate, which develops regions that remain stationary ("nodes") and regions where motion is intense ("antinodes"). Nearly 200 years ago, E. F. F. Chladni discovered that when sand was sprinkled onto vibrating plates, the different vibrational patterns were made visible as the particles danced away from the antinodes and settled on the nodes. Physical principles are rarely so graphically displayed, perhaps accounting for the popularity of this simple exercise in physics laboratories. It is this experiment for which Chladni is best known. This is the same Chladni whose pioneering work first made the study of meteorites a legitimate scientific endeavor. It is somewhat disconcerting, at least to me, that Chladni's physics experiment should have gained him more recognition than his role as the father of

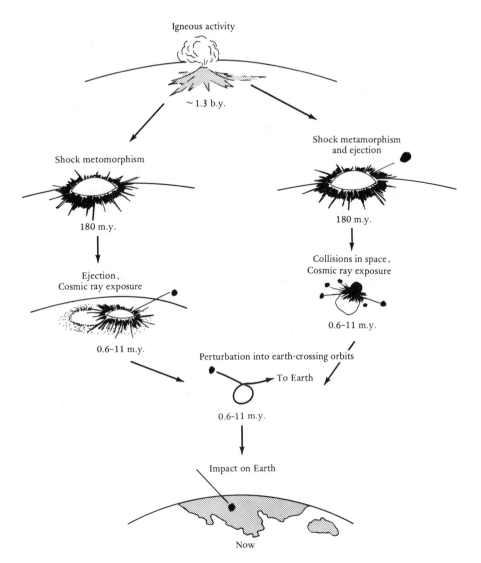

Igneous activity

~ 1.3 b.y.

Shock metomorphism

Shock metamorphism and ejection

180 m.y.

180 m.y.

Ejection, Cosmic ray exposure

Collisions in space, Cosmic ray exposure

0.6–11 m.y.

0.6–11 m.y.

Perturbation into earth-crossing orbits

To Earth

0.6-11 m.y.

Impact on Earth

Now

Fig. 8.9. The events recognized in shergottites suggest the evolutionary patterns depicted in this figure.

meteoritics, but the importance of meteorites is not always recognized, even today.

What is the value of meteorites? The trickle of interplanetary stuff that arrives on earth each year continually brings with it new and often unexpected insights into the origin and evolution of our solar system. It is probably safe to say that Chladni would be astounded at the information that has been wrested from meteorites since he first argued for their extraterrestrial origin. And the prospects for more surprises are excellent. The lifetimes for materials in earth-crossing orbits are so short that we have seen only a tiny fraction of parent-body diversity, the results of a few fairly recent collisions among asteroids or impacts into the moon and possibly Mars, and maybe a spent comet or two. The Antarctic meteorite collections sampled a population of objects orbiting at an earlier time, but these are still dominated by the same kinds of meteorites that fall today. A few unique meteorites defy classification into existing groups and must represent pieces of otherwise unsampled parent bodies. Thus, the expectation that an exciting discovery will attend each meteorite recovery is always there.

The primary purpose of this book has been to explain the scientific value of this limited meteoritic sample. Geologic inquiry requires examination of large quantities of rock in the field and laboratory before proffering any conclusions about how any part of the earth was formed. Imagine trying to characterize entire worlds, albeit small ones in many cases, from the study of only a few kilograms of sample and without benefit of field observations. It almost seems impertinent. Yet this is the task that meteoritics has set itself. It is therefore obvious why each meteorite fall or find is so important, even if it is a small ordinary chondrite. Meteorites are tantalizing gifts of new information about the solar system, just waiting to be unwrapped. We are compelled, as a consequence of our voracious need to understand our surroundings, to open these gifts and examine their contents, until such time as we can visit their parent bodies ourselves.

SUGGESTED READINGS

The topics in this chapter have not yet been treated in many nontechnical publications, and the technical literature makes for difficult reading. However, the references below contain a wealth of information for the serious reader.

GENERAL

Wood J. A. (1979) *The Solar System*, Prentice-Hall, Englewood Cliffs, NJ, 196 pp. (A nontechnical overview with an excellent chapter on celestial motions.)

Wasson J. T. and Wetherill G. W. (1979) Dynamical, chemical and iso-topic evidence regarding the formation locations of asteroids and me-teorites. In *Asteroids*, edited by T. Gehrels, University of Arizona Press, Tucson, pp. 926–974. (Technical review summarizing a tremendous amount of information bearing on the formation locations of meteo-rites.)

ASTEROID COLLISIONS

Chapman C. R. and Davis D. (1975) Asteroid collisional evolution: Evi-dence for a much larger early population. *Science* 190, 553–556. (Technical paper describing the role played by collisions in the history of the asteroid belt.)
Bogard D. D. (1979) Chronology of asteroid collisions as recorded in me-teorites. In *Asteroids*, edited by T. Gehrels, University of Arizona Press, Tucson, pp. 558–578. (Technical review summarizing the isotopic constraints on asteroid and meteoroid collisions.)

EJECTION FROM PLANETS

Melosh H. J. (1984) Impact ejection, spallation, and the origin of meteo-rites. *Icarus* 59, 234–260. (Technical paper that describes calculations of impact ejection models for planets.)

ORBITAL EVOLUTION

Greenberg R. and Scholl H. (1979) Resonances in the asteroid belt. In *Asteroids*, edited by T. Gehrels, University of Arizona Press, Tucson, pp. 310–333. (Technical paper describing orbital resonances and their role in placing meteoroids into earth-crossing orbits.)
Wetherill G. W. (1979) Apollo objects. *Scientific American* 240, 54–65. (Nontechnical paper describing the characteristics of earth-crossing asteroids.)
Wetherill G. W. (1985) Asteroidal source of ordinary chondrites. *Meteori-tics* 20, 1–22. (Technical paper in which orbital constraints on the sources of chondrites are reviewed.)

COSMIC-RAY EXPOSURE

Crabb J. and Schultz L. (1981) Cosmic ray exposure ages of ordinary chondrites and their significance for parent body stratigraphy. *Geo-chimica et Cosmochimica Acta* 45, 2151–2160. (Technical paper pre-senting data on cosmic-ray exposure of chondrites in space.)

Appendix of minerals

amphibole a class of hydrous silicate minerals with similar crystal structures, orthorhombic or monoclinic, common in terrestrial rocks but almost completely absent from meteorites.

anorthite (see **plagioclase**).

augite (see **pyroxene**).

chromite black oxide, $FeCr_2O_4$, cubic, minor constituent of iron, stony-iron, and some stony meteorites.

clay minerals (see **phyllosilicates**).

cohenite opaque carbide, bronze-colored, $(Fe,Ni,Co)_3C$, orthorhombic, minor constituent of iron meteorites.

corundum colorless oxide, Al_2O_3, hexagonal, occurs in calcium-aluminum inclusions in chondrites, sapphire and ruby are colored varieties.

diamond transparent carbon, C, dense-packed structure that forms at extreme pressures, a polymorph of graphite, produced by shock in some iron meteorites and ureilites.

diopside (see **pyroxene**).

enstatite (see **pyroxene**).

feldspar abundant class of silicate minerals, monoclinic or triclinic, includes **plagioclase** and orthoclase, $KAlSi_3O_8$, occurs mainly in achondrites and as a product of metamorphism in chondrites.

forsterite (see **olivine**).

garnet complex silicate solid solution with general formula $(Ca,Mg,Fe^{+2},Mn)_3(Al,Fe^{+3})_2(SiO_4)_3$, cubic, common in terrestrial rocks formed at high pressures.

graphite opaque carbon, C, hexagonal, sheetlike structure, a polymorph of diamond, stable at low pressures, occurs in some ureilites, ordinary chondrites, and iron meteorites.

ilmenite black oxide, $FeTiO_3$, orthorhombic, common accessory mineral in achondrites and lunar rocks.

kaersutite an amphibole, approximate composition $Ca_2Na(Mg,Fe^{+2})_4TiAl_2 Si_6O_{22}(OH)_2$, monoclinic, occurs as a minor constituent in shergottites.

kamacite (see **metal**).

magnetite black oxide, Fe_3O_4, cubic, common accessory mineral in carbonaceous chondrites, shergottites, and terrestrial rocks.

maskelynite plagioclase that has been transformed into glass (substance with no crystal structure) by shock metamorphism, occurs in shergottites and some shocked lunar and terrestrial rocks.

melilite silicate solid solution between $Ca_2Al_2SiO_7$ and $Ca_2MgSi_2O_7$, tetragonal, occurs in calcium-aluminum inclusions in chondrites.

metal (nickel-iron) alloys, kamacite has low (up to 7.5 weight percent) nickel, taenite has high (greater than 20 weight percent) nickel, cubic, major constituents of irons and stony-irons, minor constituents of many stony meteorites.

oldhamite opaque sulfide, CaS, minor constituent of enstatite chondrites.

olivine silicate solid solution between forsterite (Mg_2SiO_4) and fayalite (Fe_2SiO_4), very abundant in chondrites, pallasites, and a few achondrites.

osbornite nitride, TiN, a rare mineral in enstatite chondrites.

perovskite white oxide, $CaTiO_3$, cubic, a minor constituent of calcium-aluminum inclusions in chondrites.

phyllosilicate a broad class of hydrous silicate minerals with crystal structures in the form of stacked sheets, includes submicroscopic clay minerals with complex compositions and serpentine, $Mg_6Si_4O_{10}(OH)_8$, these occur in the matrices of carbonaceous chondrites.

pigeonite (see **pyroxene**).

plagioclase an important class of silicate minerals (feldspars) ranging in composition between anorthite ($CaAl_2Si_2O_8$) and albite ($NaAlSi_3O_8$), triclinic, a major constituent of most achondrites and lunar rocks.

pyroxene a major group of silicate minerals with similar crystal structures, complex solid solution between enstatite ($Mg_2Si_2O_6$), diopside ($CaMgSi_2O_6$), ferrosilite ($Fe_2Si_2O_6$), and hedenburgite ($CaFeSi_2O_6$), orthorhombic or monoclinic, an important constituent of chondrites, achondrites, and mesosiderites.

schreibersite gray phosphide, $(Fe,Ni)_3P$, tetragonal, a minor mineral in iron meteorites.

serpentine (see **phyllosilicate**).

sinoite Si_2N_2O, a rare mineral in enstatite chondrites.

spinel a class of oxides with the general formula

$(Mg,Fe^{+2})(Al,Fe^{+3},Cr)_2O_4$, cubic, occurs as $MgAl_2O_4$ in calcium-aluminum inclusions, chromite and magnetite are also spinels.

taenite (see **metal**).

troilite brass-colored sulfide, FeS, hexagonal, a common constituent of most classes of meteorites.

Glossary

absorption band a valley in an asteroidal reflectance spectrum produced by absorption of certain wavelengths of energy by asteroid surface materials.

accretion the process by which particles come together to form a larger mass of material.

achondrite a class of stony meteorites that crystallized from magmas; the term literally means "without chondrules," emphasizing the distinction between these and chondrites.

agglutinate a glass-bonded aggregate of broken mineral and rock fragments; a common constituent of regolith materials formed by small impacts.

albedo the percentage of incoming radiation that is reflected by a surface.

Amor asteroid an asteroid having a perihelion distance between 1.017 and 1.3 AU.

anomalous iron an iron meteorite that cannot be readily classified by either its structure or chemical composition (see **structural classification, chemical groups**).

anorthosite an igneous rock consisting primarily of plagioclase; a major constituent of the lunar highlands.

aphelion the orbital position at which the distance between the object and the sun is the greatest.

Apollo asteroid an asteroid having a perihelion distance greater than 1.017 AU and a semi-major axis greater than 1.0 AU.

aqueous alteration transformation to new assemblage of minerals caused by reactions with water at low temperatures; this process affected some carbonaceous chondrites.

asteroid a moving object of stellar appearance, without any trace of cometary activity.

asteroid belt the region between Mars and Jupiter where most asteroids are found.

astrobleme an ancient, deeply eroded crater; literally "star wound." 225

astronomical unit (AU) the distance from the earth to the sun, approximately 150 million kilometers; commonly used to express astronomical distances.

ataxite an iron meteorite with high nickel content, composed almost entirely of taenite and having no obvious structure.

Aten asteroid an asteroid having an aphelion distance of greater than 0.983 AU and a semimajor axis less than 1.0 AU.

basalt a common volcanic igneous rock consisting predominantly of pyroxene and plagioclase.

breccia a rock composed of broken rock fragments (clasts) cemented together by finer-grained material; a common product of impact processes.

calcium-aluminum inclusion a large, white inclusion, commonly irregular in shape, that occurs in carbonaceous chondrites; these are composed primarily of minerals containing calcium and aluminum and may represent condensates or refractory residues from volatilization that formed in the solar nebula.

carbonaceous chondrite a primitive class of stony meteorite, the chemical composition of which closely matches that of the sun; these are the most highly oxidized and volatile-rich chondrites.

Ceres the largest asteroid, possibly a parent body for carbonaceous chondrites.

chalcophile elements with a geochemical affinity for sulfide phases.

chassignite an achondrite consisting mostly of olivine.

chemical class (chondrites) a classification of chondrites into groups with distinct chemical compositions; commonly recognized classes are C (carbonaceous), E (enstatite), H, L, and LL (ordinary).

chemical group (irons) a classification of iron meteorites based on abundances of nickel, gallium, and germanium; groups are identified by Roman numerals and letters, as IAB, IVA.

chondrite an abundant type of stony meteorite characterized by the presence of chondrules (except in the case of C1).

chondrule millimeter-sized spherule of rapidly cooled silicate melt, found in abundance in chondritic meteorites.

clast rock fragment found in breccias.

color index the difference in magnitude between two spectral regions; in asteroid spectra this normally refers to a plot of ultraviolet-minus-blue versus blue-minus-violet.

comet a diffuse, generally unstable body of gas and solid particles that orbits the sun in a highly elliptical or parabolic orbit; comets may be short-period (periodic) or long-period.

commensurability a term applied to two bodies orbiting around the same point when the period of one is an integral multiple of that of the other.

condensation a process by which solids form directly from a gas, usually in response to lowering temperature.

cosmic abundance the abundance of some element relative to another (commonly silicon) in the sun; this is equivalent to average solar system composition and is approximately equal to the chondritic abundances of all but the most volatile elements.

cosmic dust (see **micrometeorite**).

cosmic ray exposure age the length of time a small meteoroid (meter-sized or less) was exposed to cosmic rays while orbiting in space; this is measured from the amounts of certain isotopes produced by cosmic rays (see **spallation**).

cosmic velocity the velocity of a meteoroid orbiting in space; this may be reduced by air friction during atmospheric passage.

crystallization the process of producing minerals with ordered atomic structures.

cumulate an igneous rock produced by accumulation of crystals separated from a magma through some physical process.

differentiation the process by which an initially homogeneous planetary body becomes internally stratified into regions of different compositions; this usually produces a core, mantle, and crust.

diffusion the process of permeation of any region by gradual movement and scattering of atoms in the solid state.

diogenite an achondrite composed primarily of cumulate pyroxenes; related to eucrites.

dirty-snowball model a popular model to explain the structure of comets; these may consist of water, methane, and ammonia ices mixed with particles of silicates, metal, and sulfides.

eccentricity the amount by which an elliptical orbit deviates from circularity.

electromagnetic induction a method proposed for asteroidal heating, using electric currents induced by an early solar wind.

enstatite chondrites a class of stony meteorites containing abundant enstatite and metal; these are the most highly reduced chondrites.

equilibrium a state in which a process has produced its total effect or finished its work, and therefore causes no further change.

escape velocity the velocity that any object must achieve to escape the gravitational field of its parent body.

eucrite a class of achondrites that formed as basaltic flows on their parent body; these consist mostly of plagioclase and pyroxene.

explosion crater a large depression produced by meteoroid impact.

exposure age (see **cosmic-ray exposure age**).

exsolution the separation of two previously mixed phases in the solid state; this normally produces oriented plates of one mineral within another.

fall a recovered meteorite that was observed to fall.

find a recovered meteorite that was not observed to fall.

fireball (see **meteor**).

fission track a submicroscopic trail in a crystal along which a particle produced by decay of a heavier isotope traveled; these can be used as a means of age determination.

formation in geology, any assemblage of rocks, usually stratified, that have some character (e.g., composition, age) in common.

formation interval the length of time between the origin of the solar system and the formation of meteorites.

fractional crystallization crystallization in which solid materials are physically removed from contact with the liquid from which they grew; a common process affecting most magmas.

fractionation (see **fractional crystallization**).

Fremdlinge tiny nuggets of platinum-group metals that occur in some calcium-aluminum inclusions; literally "little strangers."

fusion crust the glassy exterior of a meteorite; a melted rind that forms during atomspheric passage.

glass solid material without any crystal structure, produced by rapid cooling of magma.

half-life the time interval required for half of the remaining atoms of a certain radioactive isotope to decay; this must be known for radioactive age determinations.

hexahedrite an iron meteorite with low nickel content, consisting almost exclusively of kamacite.

highlands the ancient crust of the moon, consisting mostly of anorthosite and related rocks; these regions are distinguished by their high topography and heavy cratering.

Hirayama family a group of asteroids with similar proper orbital characteristics; these presumably represent pieces of an earlier larger asteroid that was disrupted.

howardite an achondrite breccia containing rock and mineral fragments of eucrites and diogenites.

hypervelocity impact impact of a meteoroid traveling at greater than free-fall velocity; this produces an explosion crater.

inclination the angle between an asteroid's orbit and the ecliptic plane.

ion an atom with an electrical charge, produced by loss or gain of electrons.

iron meteorite a meteorite composed primarily of iron-nickel metal (see **ataxite, hexahedrite, octahedrite**).

isochron a straight line on an isotope-evolution diagram, defined by the isotopic compositions of mineral separates; the slope of this line gives the age of the rock.

isotope one of two or more atoms having the same atomic number but different mass numbers; these can be stable or unstable (see **radionuclide**).

Kirkwood gaps voids in the asteroid belt where the orbital periods of asteroids are certain fractions of the period of Jupiter.

lithophile elements with a geochemical affinity for silicate phases.

magma molten rock material, including suspended crystals and dissolved gases, that crystallizes to form igneous rocks.

mare basalt volcanic rock of basaltic composition that occurs in the lunar maria.

maria dark, generally flat areas of the moon formerly thought to be seas.

matrix dark, fine-grained material occurring between larger objects; for example, the material between chondrules in carbonaceous chondrites and between clasts in breccias.

mesosiderite a class of stony-iron meteorites consisting of metal and fragments of igneous rocks similar to eucrites and diogenites; these formed as breccias, though many have been thoroughly recrystallized during metamorphism.

metallographic cooling rate the rate at which a meteorite cooled through a temperature of about 500°C, calculated from measured diffusion profiles of nickel in metal.

metamorphism recrystallization, in the solid state, of a rock in response to high temperature or pressure.

meteor a streak of light in the sky produced by transit of a meteoroid through the earth's atmosphere; also, the glowing meteoroid itself.

meteorite extraterrestrial material that survives passage through the atmosphere and reaches the earth's surface as a recoverable object.

meteoroid a small object orbiting the sun in the vicinity of the earth.

micrometeorite small particles, possibly of cometary material, that generally melt or are vaporized during atmospheric passage;

these have been trapped in their unaltered states in the upper atmosphere by high-altitude aircraft.

nakhlite a class of achondrites composed primarily of cumulus augite.

nebula (see **solar nebula**).

Neumann lines twin boundaries in metal grains of iron meteorites formed by shock deformation.

octahedrite an iron meteorite of intermediate nickel content, containing both kamacite and taenite in a Widmanstatten pattern.

onion-shell model a proposed model for chondrite parent bodies in which various petrologic types (metamorphic grades) are arranged concentrically within the body.

Oort cloud a massive swarm of comets thought to surround the solar system in a spherical volume extending out as far as 50,000 AU.

orbit the path, usually elliptical, followed by one object revolving around another.

ordinary chondrite the most common class of stony meteorites; this group actually consists of several chemical classes – H, L, and LL.

organic matter compounds of carbon, hydrogen, and commonly oxygen; these typically form large, complex molecules and occur in chondritic meteorites.

oriented meteorite a meteorite of conical shape formed during transit through the earth's atmosphere.

paleomagnetism the magnetism of a rock acquired in a past geologic age.

pallasite a class of stony-iron meteorites consisting of metal and isolated crystals of olivine.

parent body objects of asteroidal size or larger from which meteorites were derived.

partition coefficient the ratio of the concentration of an element in a solid relative to its concentration in a liquid with which it is in equilibrium.

peridotite an ultramafic rock composed primarily of olivine and pyroxene; probably the dominant rock type in the mantles of the earth and other bodies.

period the time required for an orbiting object to make one complete revolution.

perturbation any disturbance or minor sinuosity imposed on an otherwise elliptical orbit; this may be caused by gravitational attraction of a third body.

petrologic type a relative scale from 1 to 6, reflecting increasing metamorphic recrystallization in chondrites.

phase diagram a diagram showing the stability fields of various minerals in terms of temperature, pressure, or composition.

plutonic igneous rocks that crystallize in the interior of a planet or asteroid.

polarimetry the study of how light is polarized during reflection from an asteroidal surface.

polymorph one of several crystallographic forms taken by the same chemical compound or element; for example, graphite and diamond are polymorphs of carbon.

protoplanet a small body that will ultimately grow to form a planet by accretion.

radiometry the study of the thermal brightness of asteroids, measured from infrared radiation.

radionuclide an unstable (radioactive) isotope.

raisin-bread model a proposed model for iron meteorite parent bodies, in which masses of iron are scattered throughout the interiors rather than forming large central cores.

rare earth elements lanthanide series of the Periodic Table; closely related elements with large ionic radii that are useful in geochemical modeling.

reflectance spectra (see **spectrophotometry**)

regmaglypt depression resembling a thumbprint that is produced on the surfaces of some meteorites during atmospheric transit.

regolith a layer of fragmental, incoherent rocky debris that nearly everywhere forms the surface terrain; it is produced by repeated meteorite impact.

regolith breccia compacted and lithified regolith material.

resonance the selective response of any oscillating system to some external stimulus of the same frequency as that of the system; resonance perturbs asteroid orbits.

rubble-pile model a proposed model for chondrite parent bodies in which large asteroids form by accretion of smaller, already metamorphosed planetesimals; thus, petrologic types are distributed randomly within parent bodies.

semi-major axis the long axis of an ellipse; one characteristic of orbits.

shergottite a class of achondrites of basaltic composition, consisting mostly of plagioclase and pyroxene; all are shock-metamorphosed and have young crystallization ages.

shock metamorphism the effects of impact into a target rock;

these include transformation of minerals into new crystal structures, brecciation, melting, and other modifications.

siderophile elements with a geochemical affinity for metallic iron.

silicate inclusions pieces of silicate rock contained in iron meteorites.

solar nebula the primitive disk-shaped cloud of gas and dust from which all bodies in the solar system originated.

solar wind the radial outflow of energetic charged particles from the sun; these may be implanted in regolith materials.

spallation the production of new isotopes through irradiation by cosmic rays.

speckle interferometry a technique that eliminates some of the distorting effects of the earth's atmosphere and allows better telescopic resolution of asteroids.

spectrophotometry the study of the spectrum of sunlight reflected from the surface of an asteroidal body.

stony-iron meteorite a class of meteorites consisting of approximately equal parts silicate minerals or rock and iron-nickel metal (see **mesosiderite, pallasite**).

stony meteorite the most common class of meteorites, consisting mostly of silicate minerals (see **chondrite, achondrite**).

strewn field a generally elliptical pattern of distribution of recovered meteorites, formed when a meteoroid is fragmented during atmospheric transit.

striations parallel lines or tiny grooves on the surface of a crystal.

structural classification (irons) a classification of irons by their observable structural features; recognized groups are ataxites, hexahedrites, and octahedrites; the latter are further distinguished by the widths of kamacite bands.

supernova the explosion of a star; this results in the formation of new elements and in their dispersal.

terrestrial age the length of time a meteorite has resided on earth, measured from the decay of radionuclides produced by cosmic-ray exposure in space.

ureilite a class of achondrites consisting mostly of cumulus pyroxenes set in a carbon-rich matrix; diamonds occur in shock-metamorphosed ureilites.

Vesta the fourth largest asteroid, possibly the parent body for eucrites, diogenites, and howardites.

volatile element an easily volatilized element that condenses from a gas at low temperature.

volcanic an igneous rock that formed by solidification of magma that erupted onto the surface of a planet or asteroid.

Widmanstatten pattern a regular geometric intergrowth of kamacite plates within taenite that occurs in some iron meteorites (see **octahedrite**).

Index

abundances
  cosmic, of elements, 41–42
  falls and finds, 6
  meteorite types, 6
achondrites
  abundance, 6
  ages, 110, 119, 144–145, 218
  characteristics, 6
  chemical composition, 108–112,
    117–118, 126
  classification, 107–108, 113–115
  eucrite association, 107–113, 138–
    141, 181
  lunar, 120–122
  mineralogy, 108, 113
  shergottite association, 113–120,
    146, 149–151
  shock, 118–119, 124, 132
  ureilites, 123–124
ages
  cosmic-ray exposure, 91, 211–214
  fission track, 40
  formation interval, 40
  radiometric, 37–38, 54, 57, 110,
    119, 176
  terrestrial, 28, 214–215
  volcanic activity, 144
agglutinates, 91, 122
albedo, 73, 77, 192
Antarctica, 23–28, 120, 157
Appendix of minerals, 222–224
asteroids
  belt, 69, 78–79, 194, 203–204
  compositional types, 77, 153, 192–
    194
  differentiated, 185–188, 197
  disruption, 196–197, 202
  earth-crossing, 71, 208–211
  887 Alinda, 71, 75
  1862 Apollo, 211
  44 Nysa, 197
  4 Vesta, 75, 140–142, 193–194, 210
  433 Eros, 71
  heating, 82–85, 137–138
  Hirayama families, 195–197
  meteorite parent bodies, 67, 141–
    142, 152, 192–194
  models, 84, 86, 139, 187

1915 Quetzalcoatl, 71, 209–210
1 Ceres, 29, 69, 75, 77
158 Koronis, 196
135 Hertha, 197
  regolith, 88–92, 142
  size distribution, 80, 142, 194
24 Themis, 196
2100 Ra-Shalom, 211
2201 Oljato, 94, 211
2 Pallas, 29, 142, 179
ataxite, *see* iron meteorites, classification
atmospheric effects, 13–15

Biot, J. B., 4
breccias, 55–56, 81, 88, 108, 135, 181

chassignite, *see* achondrites, shergottite
  association
Chladni, E. F. F., 3–4, 28, 218
chondrites
  abundance, 6
  accretion, 46
  ages, 38–40, 54, 57, 212, 215–217
  aqueous alteration, 52–55
  calcium-aluminum inclusions, 49,
    58–61
  carbonaceous, 43, 77, 94, 124
  characteristics, 6
  chemical composition, 42, 58, 82
  chondrules, 35, 48–49, 58
  classification, 43, 53, 55
  components of, 46–49
  cooling rates, 85–86
  enstatite, 43, 77
  formation, 63
  metamorphism, 50, 217
  ordinary, 43, 77
  oxidation state, 44
  petrologic types, 52–53
  shock, 55
comets, 67, 92–95
condensation, 58
cores
  asteroid, 185–188, 193
  composition, 159–160
  earth, 157–158
cosmic dust, *see* micrometeorites

cosmic rays, 91, 211–214
craters
    astrobleme, 19
    explosion, 18, 45
    lunar, 145
    Meteor Crater (Arizona), 20–22
    process, 20

de Bournon, J. L., 4
differentiation, 46
diogenite, *see* achondrites, eucrite association

eucrite, 107–113, 138–141, 181
exsolution, 139

falls, 6, 13, 27
finds, 6, 23
fireballs, 4, 14, 68

Goldschmidt, V. M., 159

hexahedrite, *see* iron meteorites, classification
historical
    falls, 1–3
    hypotheses of meteorite origin, 3–5, 48–49
    meteoritic research, 4–5
Howard, E. C., 4
howardite, *see* achondrites, eucrite association

impacts
    asteroidal, 139, 202
    extinctions, 22, 160–161
    hypervelocity, 18–19, 149, 207
    lunar, 135, 145
    Tunguska event, 95–96
iron meteorites
    abundance, 6
    ages, 176, 213
    characteristics, 5
    chemical composition, 160, 170, 172
    classification, 165–171
    cooling rates, 174–175, 186
    mineralogy, 161–165
    shock, 165–166, 177–178
    silicate inclusions, 176–177
    Widmanstatten pattern, 163–164, 166, 168, 173–175, 185
isotopes (*see also* ages)
    extinct, 40, 61, 83, 137
    fractionation, 58
    half-life, 36
    isochron, 37
    oxygen, 58–61, 81–82, 125–126, 176–178
    radioactive, 35, 110, 119, 145, 176
    spallation, 211

Kirkwood gaps, 204

magma
    crystallization, 103–104, 106
    fractional crystallization, 104, 106–107, 171, 173
    ocean, 132
    origin, 102–103
mesosiderite, 180–182, 188
metamorphism, 50, 217
meteorites (*see also* achondrites, chondrites, iron meteorites; stony-iron meteorites)
    Antarctic, 23, 120, 215
    classification, 5–6
    definition, 5
    fusion crust, 9, 17
    mineralogy, 11, 13
    rate of fall, 9
    shapes, 9
    sizes, 7–8
meteoroids, 5, 14
meteors, 5
micrometeorites, 8, 96
minerals, 11, 13, 59, 104–105, 107–108, 222–224
moon
    Apollo program, 129, 136
    highlands, 130–133
    historical, 30
    maria, 133–134
    meteorites, *see* achondrites, lunar
    regolith, 88, 135

nakhlite, *see* achondrites, shergottite association
names of meteorites
    Allan Hills A76005 (Antarctica), 100
    Allan Hills A81005 (Antarctica), 121, 125, 136–137
    Allan Hills A81013 (Antarctica), 12
    Benld (Illinois), 18
    Bruno (Canada), 11
    Calico Rock (Arkansas), 166
    Canyon Diablo (Arizona), 20
    Cape York (Greenland), 8
    Derrick Peak A78009 (Antarctica), 156
    Dhajala (India), 69–70
    Drum Mountains (Utah), 175
    Eagle Station (Kentucky), 180
    Elephant Moraine A79001 (Antarctica), 149–150
    Ensisheim (France), 1, 31
    Farmington (Kansas), 60–70
    Guarena (Spain), 51
    Hoba (Namibia), 7, 169
    Ibitira (Brazil), 110
    Innisfree (Canada), 69–70
    Jilin (China), 8, 17

Johnston (Colorado), 109
Juashiki (Japan), 16
Kenna (New Mexico), 124
L'Aigle (France), 4
Lost City (Oklahoma), 69–70
Mount Stirling (Australia), 167
Nakhla (Egypt), 16, 116
Nakhom Pathom (Thailand), 56
Novo Urei (Russia), 152–153
Orgueil (France), 54, 62
Pasamonte (New Mexico), 109
Pitts (Georgia), 177
Pribram (Czechoslovakia), 67, 70
Punjab (India), 179
Reckling Peak A79015 (Antarctica), 181
Richardton (North Dakota), 39
Salta (Argentina), 180
Shergotty (India), 115
Stannern (Czechoslovakia), 138
Suga Jinja (Japan), 1
Sylacauga (Alabama), 16
Tieschitz (Czechoslovakia), 38
Vigarano (Italy), 47
Waingaromia (New Zealand), 168
Weston (Connecticut), 5
Wethersfield (Connecticut), 13–17
Willamette (Oregon), 8
Wold Cottage (England), 4
Zagami (Nigeria), 116
nebula, 58, 137

octahedrite, *see* iron meteorites, classification
Olbers, H. W., 29, 71
Oort cloud, 92
orbits
asteroid, 71, 208–210
calculations, 120
comet, 93
elements of, 202–203
meteoroid, 14, 67–70
perturbations, 203–205, 217–219
resonances, 205
organic matter, 61–63
oriented meteorite, 9

Pallas, P. S., 179
pallasite, 179–180, 188–190
parent bodies (*see also* asteroids, comets, planets)
chondrite, 67, 196
eucrite, 137–142
historical, 28–30
iron meteorite, 185–194
moon, 129–137
shergottite, 143–144, 147–152, 206–208
stony-iron meteorite, 193
ureilite, 152–154

phase diagram, 161–163
Piazzi, G., 29
planets
Jupiter, 72, 145, 204, 206
Mars, 72, 143–144, 147–152, 206
protoplanets, 45
Venus, 101, 144
polarimetry, 78, 142
polymorphs, 119

radiometry, 78, 142
rare earth elements, 106–107, 112, 118, 124, 134, 146
reflectivity, *see* albedo; spectra, reflectance
regmaglypts, 9, 14
regolith, *see* asteroids, regolith
rocks, igneous
anorthosite, 121, 132
basalt, 101, 107, 113, 117, 121, 133
cumulates, 104–105, 113–114, 123, 147, 152
peridotite, 113, 138

satellites (*see also* moon)
Diemos, 71, 202
Io, 143, 145
Phobos, 71–72, 90, 202
shergottite, *see* achondrites, shergottite association
shock, *see under* achondrites; chondrites; iron meteorites
solar nebula, *see* nebula
Sorby, H. C., 48
spectra, reflectance
asteroid, 75–77, 140–142, 193, 195–196
meteorite, 76, 140, 153, 191–193
spectrophotometry, 73
stony-iron meteorites
abundance, 6
characteristics, 6
mesosiderites, 180–182, 188
pallasites, 179–180, 188–190
strewn field, 16–17
supernova, 61

thunderstones, 3
Titius-Bode law, 29
trace elements, *see* rare earth elements

Urey, H. C., 30, 62

velocity
asteroid, 201
cosmic, 14, 17
escape, 202, 207–208
meteoroid, 14–15
volatile elements, 42–44, 84